FLORA OF TROPICAL EAST AFRICA

COMBRETACEAE

G. E. WICKENS*

Trees, shrubs, shrublets or climbers, very rarely subherbaceous. Indumentum of unicellular compartmented (very rarely non-compartmented) hairs, multicellular stalked glands and multicellular scales (in which the head consists of a multicellular plate only one cell thick). Leaves opposite, verticillate, spiral or alternate, exstipulate, simple and almost always entire (very rarely crenulate). Flowers ♀ or ☿ and ♂ in the same inflorescence, usually 4–5-merous, rarely slightly zygomorphic, in axillary or extra-axillary spikes or racemes or in terminal or axillary panicles. Receptacle (calyx-tube or hypanthium) usually in 2 distinct parts, the lower receptacle surrounding and adnate to the inferior (except in *Strephonema* from West Africa) ovary and the upper receptacle usually produced beyond it to form a short or long tube terminating in the sepals (calyx-lobes), the latter sometimes scarcely developed. Sepals 4 or 5 (rarely 6 or 8) or almost absent, sometimes accrescent (*Calycopteris* in Asia). Petals 4 or 5, conspicuous to very small, or absent, usually inserted near the mouth of the upper receptacle. Stamens usually twice as many as the sepals or petals (occasionally the same number, but not in E. Africa), biseriate or more rarely uniseriate, exserted or included in the upper receptacle; anthers versatile (rarely adnate to the filaments but not in E. Africa). Disk intrastaminal, hairy or glabrous, sometimes inconspicuous or absent. Style usually free (attached to the upper receptacle for part of its length in *Quisqualis* and a very few species of *Combretum*); ovary inferior (semi-inferior in *Strephonema*), 1-locular with usually 2 (up to 6) pendulous anatropous ovules of which only 1 usually develops. Fruit very variable in size and shape, fleshy or dry, stipitate or sessile, usually indehiscent, often variously winged or ridged, 1-seeded. Seeds without albumen. Cotyledons 2 (rarely 3), occasionally with their petioles connate almost to the apex.

Widespread in the tropics and subtropics, with 20 genera (according to Exell & Stace in Bol. Soc. Brot., sér. 2, 40: 5–25 (1966)), 11 of which occur in tropical Africa, and about 500 species.

The compartmented hairs (requiring microscopic examination) known also as " combretaceous hairs " occur in all E. African species (unless the specimen is entirely glabrous) and are known, apart from the *Combretaceae*, only in the *Cistaceae* and a few species of *Myrtaceae* so that they give an almost certain method of identifying the family even from fragmentary material (see C. A. Stace in J.L.S. 59: 233–234, figs. 1–12 (1965)).

NOTE. The reference to *Anogeissus* in Burtt Davy, Check-lists Brit. Emp. 1, Uganda: 35 (1935) refers to a collection made by Grant in the Sudan. The genus has not been recorded from the Flora area.

Conocarpus lancifolius Engl. & Diels, a tall tree from Somali Republic, is cultivated in K7 near Mombasa, where it is grown on land being reclaimed after excavations of coral limestone for cement manufacture.

* The writer is deeply grateful to Dr. A. W. Exell for his constant encouragement and advice during the preparation of this account.

Lower receptacle without adnate bracteoles; not
   mangroves :
  Petals present (except possibly in aberrant flowers);
    leaves usually opposite or verticillate :
    Flowers all ♀; scales or stalked glands (usually
      microscopic) present :
      Style not adnate to the upper receptacle (except
        in two species not in E. Africa and in these
        the stamens are exserted); scales or stalked
        glands (usually microscopic) always
        present . . . . . .   **1. Combretum**
      Style adnate to the upper receptacle for part
        of its length; stamens not exserted beyond
        the petals; scales absent; microscopic
        stalked glands present . . . .   **2. Quisqualis**
    Flowers ♀ and ♂ in the same inflorescence;
      neither scales nor microscopic stalked glands
      present . . . .   **3. Pteleopsis**
  Petals absent; flowers usually ♀ and ♂; neither
    scales nor microscopic stalked glands present;
    leaves usually spirally arranged or alternate,
    sometimes in fascicles on spur shoots, some-
    times (in *Conocarpus*) subopposite :
    Flowers in spikes or spike-like racemes, never
      aggregated into tight cones . . .   **4. Terminalia**
    Flowers and small fruits aggregated into tight
      cones . . . .   **Conocarpus**
                                    (see note above)

Lower receptacle with 2 adnate bracteoles; mangroves . . . .   **5. Lumnitzera**

1. COMBRETUM

Loefl., Iter Hispan.: 308 (1758), *nom. conserv.* For full synonymy see Exell
        in J.B. 69: 116 (1931)

   Trees, shrubs, shrublets or woody climbers, very rarely subherbaceous;
scales or microscopic (and sometimes macroscopic) stalked glands present.
Leaves opposite, verticillate or rarely alternate, usually petiolate, almost
always entire; petiole sometimes persisting (especially in climbers), forming
a ± hooked spine. Flowers ♀ (in Flora area), regular or slightly zygomorphic
(not in Flora area), 4–5-merous, in elongated or subcapitate axillary or extra-
axillary spikes or racemes or in terminal or terminal and axillary, often leafy,
panicles. Receptacle usually clearly divided into a lower part (lower receptacle) surrounding and adnate to the ovary, and an upper part (upper
receptacle) varying from patelliform to infundibuliform and itself sometimes
visibly differentiated into a lower part containing the disk (when present) and
an often more expanded upper part. Sepals 4 or 5 (rarely more), deltate to ±
subulate or filiform, sometimes scarcely developed. Petals usually 4 or 5 in
Flora area (rarely absent in aberrant specimens), small and inconspicuous or
showy and exceeding the sepals. Stamens twice as many as the petals,
inserted in 1 or 2 series inside the upper receptacle and usually exserted.
Disk glabrous or hairy, with or without a free margin, sometimes inconspicuous or absent. Style free (in E. African species); stigma sometimes ±
expanded; ovary completely inferior. Fruit 4–5-winged, -ridged or -angled,
sessile or stipitate, indehiscent or rarely tardily dehiscent; pericarp usually

thin and papery, sometimes leathery, more rarely fleshy. Cotyledons various.

About 250 species, throughout the tropics (except Australia and Pacific Is.) and extending into the subtropics.

Three subgenera are recognised by Exell & Stace in Bol. Soc. Brot., sér. 2, 40: 10 (1966); they are subgen. *Combretum, Cacoucia* (Aubl.) Exell & Stace and the monotypic Asian subgen. *Apetalanthum* Exell & Stace. On a worldwide scale the subgenera *Combretum* and *Cacoucia* are separable with certainty only on the character of the presence of either scales (subgen. *Combretum*) or microscopic stalked glandular hairs (subgen. *Cacoucia*) and, except when scales are evidently visible with the naked eye or a × 20 lens, microscopic examination is necessary. When in doubt, however, it is possible within the Flora area (though not everywhere in Africa) to separate flowering material into the two subgenera fairly satisfactorily without microscopic examination by use of the second somewhat artificial key. An artificial key for the identification of material with fruits and leaves but no flowers is given below.

KEY TO SUBGENERA

Scales present though sometimes inconspicuous or
    hidden by the indumentum; microscopic stalked
    glands absent; flowers usually 4-merous .   . 1. subgen.
          Combretum (p. 9)

Scales absent; microscopic stalked glands present;
    flowers 4-merous or 5-merous .   .   . 2. subgen.
          Cacoucia (p. 48)

KEY TO SUBGENERA WITHOUT USE OF SCALE CHARACTERS

Flowers and fruits 4-merous:
  Petals bright red .   .   .   .   .   . subgen.
          Cacoucia (p. 48)
  Petals white, yellowish or pinkish:
    Petals 4 mm. long or longer:
      Scales absent; flowers in handsome spikes . subgen. *Cacoucia*
      Scales conspicuous .   .   .   . subgen. *Combretum*
          (p. 9)
    Petals less than 3·5 mm. long:
      Scales absent; leaves pellucid-punctate. . subgen. *Cacoucia*
          sect. *Mucronata*
          (p. 49)
      Scales present but sometimes inconspicuous . subgen. *Combretum*
Flowers and fruits 5-merous .   .   .   . subgen. *Cacoucia*

NOTE. Most, but not all, climbers belong to subgen. *Cacoucia*.

ARTIFICIAL KEY TO SPECIES BASED ON FRUIT AND VEGETATIVE FEATURES

NOTE. Specific identification is not always easy at the fruiting stage and some of the distinctions used in the following key are not entirely invariable, so that trial of alternative leads and close attention to the specific descriptions may sometimes be necessary. Fruits of 19, *C. coriifolium*, 23, *C. chionanthoides* and 45, *C. sp. B*, are unknown so these species are not included. Fruits of most species are illustrated in figs. 4 (p. 15), 5 (p. 16) and 7 (p. 51). Length of the fruit, following the usual convention in this family, does not include the length of the stipe. A × 20 lens is necessary for observing the scale characters.

1. Fruit broadly winged (wing-width equal to or
    greater than body-width) 2
  Fruit angled or narrowly winged (wing-width
    less than body-width) 41

2. Scales (sometimes sparse) present on the fruit
     and leaves 3
   Scales absent 26
3. Scales on leaves immersed or medially im-
     pressed, relatively large (100–300 μ in
     diameter) or if smaller leaving pock-marks
     of the same order of size 4
   Scales not immersed and if impressed much
     smaller and the impression not readily
     visible with a × 10 lens, not leaving pock-
     marks 13
4. Fruits ± 4–5·5 cm. long (fig. 5/25, p. 16); leaves
     elliptic, rather densely lepidote with almost
     wholly immersed scales, with ± 7–10 pairs
     of lateral nerves . . . . . 25. *C. xanthothyrsum*
   Fruits less than 3 cm. long 5
5. Infructescences 1–3 cm. long, mostly on short ±
     divaricate lateral branches; leaf-scales
     (fig. 2/20, p. 13) mostly less than 120 μ in
     diameter, with slightly wavy margins; fruit
     (fig. 5/20, p. 16) ± 1·5–2·5 cm. long, the
     body conspicuously covered with irregu-
     larly shaped red-brown scales; shrub with
     second year bark ± in fibrous shreds . 20. *C. hereroense*
   Infructescences not on short shoots or if so
     (3, *C. contractum*) longer with larger scales
     and golden lepidote fruits 6
6. Fruits in heads, ovate in outline, pointed (fig.
     5/24, p. 16); leaves ± oblong-oblanceolate
     with undulate margins and ± 12–16 pairs
     of lateral nerves . . . . . 24. *C. capituliflorum*
   Fruits on elongate rhachides (except 44, *C. sp.*
     A), oblong to transversely elliptic, apically
     emarginate; leaves with ± flat or incurved
     margins and 3–10 pairs of lateral nerves 7
7. Scales 40–55(–70) μ in diameter, not readily seen
     with a × 10 lens, only the pock-marks con-
     spicuous; fruits ± oblong-elliptic, ± 2·4–
     2·8 cm. long, inconspicuously lepidote;
     usually climbers in E. Africa 8
   Scales mostly 120–300 μ in diameter; fruits
     mostly smaller and if exceeding 2 cm. sub-
     circular to transversely elliptic in outline,
     conspicuously lepidote; trees or shrubs,
     sometimes scandent 9
8. Stipes of mature fruits 2–2·5 mm. long; leaves
     usually obovate-elliptic, shortly and
     bluntly acuminate. . . . . 5. *C. tanaense*
   Stipes of mature fruits 5–8 mm. long; leaves
     elliptic with a more attenuate and sharp
     acumen . . . . . . 6. *C. umbricola*
9. Leaves ± rounded at the apex, with densely
     arranged to contiguous scales on both
     surfaces . . . . . . 4. *C. imberbe*
   Leaves ± acuminate, not so densely lepidote 10

10. Rhachides of infructescences less than 1 cm.
    long; leaves coriaceous with distinctly
    raised lateral nerves beneath .   .   .   44. *C. sp. A*
    Rhachides of infructescences mostly 3–10 cm.
    long; leaves (in E. Africa) membranous or
    sometimes subcoriaceous but with the fine
    lateral nerves only slightly raised beneath   .   .   .   11
11. Fruits (fig. 4/3, p. 15) oblong in outline, with
    apical peg absent or less than 0·5 mm. long,
    golden lepidote; leaves mostly crowded on
    short shoots .   .   .   .   .   .   3. *C. contractum*
    Fruits subcircular to transversely elliptic in
    outline, rufous or golden lepidote; leaves
    not crowded on short shoots   12
12. Apical peg of fruit 0·5–2·5 mm. long; petioles
    rarely more than 6 mm. long in E. Africa   1. *C. celastroides*
    Apical peg of fruit 0–0·5 mm. long; petioles
    of mature leaves 10–25 mm. long   .   .   2. *C. padoides*
13. Scandent shrubs or lianes; fruits 2–2·8 cm. long,
    glutinous   14
    Trees or shrubs, not scandent   16
14. Leaves with numerous small but readily visible
    scales, ciliate; fruits in elongate unbranched
    axillary spikes   .   .   .   .   .   16. *C. acutifolium*
    Leaves with minute inconspicuous scales,
    glabrous on the margins; fruits in panicles
    or pseudopanicles (upper leaves partially
    suppressed)   15
15. Fruits (fig. 4/6, p. 15) broadly elliptic in outline,
    with a 1–2 mm. long apical peg, in terminal
    and axillary nearly sessile spikes   .   6. *C. umbricola*
    Fruits (fig. 5/18, p. 16) subcircular in outline,
    without an apical peg, in spikes with ± 1
    cm. long peduncles, axillary or aggregated
    into pseudopanicles by suppression of
    upper leaves   .   .   .   .   .   18. *C. fuscum*
16. Rhachides of infructescences 1–2(–3) cm. long.   .   .   .   17
    Rhachides of infructescences mostly much
    longer   19
17. Leaves conspicuously lepidote or hairy, the
    scales with wavy margins, often giving a
    scurfy appearance; fruit (fig. 5/20, p. 16)
    1·5–2·5(–3·5) cm. long, with a stipe up to
    11 mm. long; many branches abbreviated
    and divaricate   .   .   .   .   .   20. *C. hereroense*
    Leaves with minute scales, glabrous except
    sometimes on the nerves; mature fruits
    more than 3 cm. long, with stipes longer
    than 1 cm.   18
18. Fruits (fig. 4/7, p. 15) glutinous when young,
    not hairy, indented at the base of the wings;
    leaf-blades mostly more than 5 cm. long;
    branches not divaricate .   .   .   .   7. *C. schumannii*
    Fruits (fig. 4/8) tomentellous when young; wings
    decurrent to the stipe; leaf-blades 2·5–5
    cm. long; branches divaricate.   .   .   8. *C. gillettianum*

19. Fruits with a somewhat " metallic " appearance,
    the body with few to many relatively large
    red-brown scales   20
    Fruits without a " metallic " appearance   21
20. Fruits (fig. 4/9, p. 15) ovate in outline, the
    body hoary with few scattered red-brown
    scales; leaf-blades elliptic, up to 8 cm.
    long, papery, glabrous, pustulate, with
    only scattered scales   .   .   .   .   9. *C. tenuipetiolatum*
    Fruits variable in shape, 2–5·9 cm. long, with
    many red-brown scales on the body (some-
    times concealed by the indumentum); leaf-
    blades variable but usually larger and
    thicker textured, always with a conspicu-
    ous covering of hairs or scales beneath   .   10. *C. collinum*
21. Fruits, (fig. 4/13, p. 15), (3–)5–6·5(–10) cm.
    long, with straw-coloured to light brown
    sparsely lepidote wings, not or only
    slightly glutinous; leaves hairy to glab-
    rescent with somewhat raised nervation
    beneath   .   .   .   .   .   13. *C. zeyheri*
    Fruits up to 3·5 cm. long   22
22. Fruits and leaves not glutinous; scales (fig.
    2/14, p. 13) ± (75–)90–120(–130) μ in
    diameter, opaque; fruits 1·3–2·5 cm.
    long   .   .   .   .   .   .   14. *C. molle*
    Fruits and usually the young leaves glutinous;
    scales mostly smaller, transparent; fruits
    often larger   23
23. Bark peeling from previous season's branches in
    dark cylindrical or hemicylindrical pieces,
    leaving a powdery cinnamon-red surface;
    leaves large, rounded or retuse at apex,
    with prominent venation beneath   .   17. *C. psidioides*
    Bark flaking in small pieces or shredding into
    fibrous strips; leaves generally acute or
    obtuse at apex   24
24. Leaves opposite, with the venation usually only
    slightly raised beneath; lateral nerves
    ascending, looped at a level between the in-
    sertion of the second or third nerve above;
    bark of younger branches shredding in
    fibrous strips or threads; fruits (fig.
    5/15, p. 16) up to 3 cm. long, less than 1·5
    times as long as broad   .   .   .   .   15. *C. apiculatum*
    Leaves opposite or some often 3(–4)-verticillate,
    with venation of mature leaves prominently
    raised beneath; lateral nerves more spread-
    ing, looped at a level between the insertion
    of the first and second nerve above; bark
    of younger branches coming away irregu-
    larly in small pieces and strips; fruits
    (fig. 4/11, 12, p. 15) often more than 3 cm.
    long, often 1·5–2 times as long as broad   25
25. Fruits glabrous apart from the rather incon-
    spicuous scales; leaves glabrescent .   .   11. *C. fragrans*

Fruits pubescent when young; leaves densely
tomentose on the prominent reticulation
beneath   .   .   .   .   .   .   12. *C. schweinfurthii*

26. Fruits 4-winged   27
 Fruits 5(-6)-winged   32

27. Fruits (fig. 7/31, 32, p. 51) ± obovate, 1·8–2·5
 cm. long, in clusters on several cm. long
 side branches of a terminal panicle, the
 bracts of which are white or coloured
 (though mostly fallen at fruiting stage) .   .   .   28
 Fruits on an elongate rhachis or the lateral
 panicle-branches short, without white or
 coloured bracts*   29

28. Fruits glabrous .   .   .   .   .   31. *C. racemosum*
 Fruits puberulous   .   .   .   .   32. *C. cinereopetalum*

29. Fruits (fig. 5/28, p. 16) 1–1.8 cm. long with a
 ± 1 mm. long stipe and thin papery wings,
 borne on extensive rufous pubescent
 panicles   .   .   .   .   .   28. *C. mucronatum*
 Fruits larger   30

30. Rhachides of infructescences 4–8 cm. long,
 shortly pedunculate in axils of mature or
 fallen leaves   31
 Rhachides of infructescences mostly short, up
 to 1 cm. long, the upper leaves normally
 progressively reduced to give the appearance of panicles   36

31. Fruits velutinous, apparently always 4-winged
 (**T**8)   .   .   .   .   .   .   37. *C. andradae*
 Fruits inconspicuously pubescent, rarely 4-
 winged (widespread inland)   .   .   .   40. *C. mossambicense*

32. Fruits (fig. 7/33, p. 51) subsessile, hairy, in
 heads .   .   .   .   .   .   33. *C. obovatum*
 Fruits stipitate or glabrous   33

33. Infructescence a false raceme with the ultimate
 1–3 mm. long pedicel-like branches
 persistent and crowded on the 3–10(–17)
 cm. long tomentose rhachides, the leafy
 bracts small and mostly fallen at this
 stage .   .   .   .   .   .   36. *C. rhodanthum*
 Infructescences simply racemose or if aggre-
 gated into panicles without such pedicel-
 like ultimate branches   34

34. Fruits (fig. 7/38, p. 51) mostly less than 2 cm.
 long with a slender stipe $\frac{1}{3}$–$\frac{1}{2}$ as long
 as the fruit; infructescences axillary,
 numerous, short, mostly on relatively
 short lateral branches; rambling shrub
 with many persistent recurved spine-like
 petioles   .   .   .   .   .   38. *C. aculeatum*
 Fruits mostly larger with a relatively short
 stipe   35

35. Rhachides of infructescences mostly short, up
 to 1–2 cm. long, the upper leaves normally

* 6, *C. umbricola* has minute scales and if these are not observed it might key to this
dichotomy but the fruits are ellipsoid and the panicle-branches well developed without
white or coloured bracts.

Leaves narrowly oblong-elliptic to obovate, pubescent to tomentose; fruit (fig. 5/27) 2–3 cm. long, angled or shallowly winged . . . . . 27. *C. exalatum*

46. Fruit (fig. 7/34, p. 51) narrowly obovate in outline, 2–4·5 cm. long; infructescences capitate; petioles 5–15 mm. long, infrequently forming spines . . . . . 34. *C. pentagonum*

Fruit (fig. 7/43) ellipsoid, ± 2·5 cm. long; infructescences spicate; petioles mostly 4–6 mm. long, rarely longer, mostly persisting as spines . . . . . 43. *C. constrictum*

Subgenus **COMBRETUM**

Scales present (sometimes inconspicuous or concealed by the indumentum). Flowers and fruits normally 4-merous.

NOTE. The identification of the sections within subgenus *Combretum* is not easy. Microscopic examination of the scales is the simplest and most accurate method, but unfortunately it is not always very practical for field botanists.

KEY TO THE SECTIONS

Upper receptacle (fig. 3/1, p. 14) little developed, almost flat; disk conspicuously visible, well developed; petals glabrous, linear-elliptic; stamens 1-seriate; scales (fig. 1/1–4, p. 12) conspicuous, large, usually at least (100–)150 μ in diameter, divided by many radial and tangential walls . . . . . 1. *Hypocrateropsis*, p. 11

Upper receptacle cupuliform to infundibuliform; disk usually concealed within the upper receptacle and usually not visible in dried specimens without opening the flower; petals glabrous or hairy, of various shapes; scales various (sometimes inconspicuous), if over 150 μ in diameter either conspicuously wavy or scalloped at the margin or divided almost solely by radial walls:

Petals ciliate at the apex, 0·5–1·1(–1·5) mm. long, obtriangular or obovate, often rather inconspicuous:

Stamens 1-seriate; fruit broad-winged . . 7. *Ciliatipetala*, p. 32

Stamens 2-seriate; fruit with very narrow (1 mm. wide) wings (*C. illairii, C. capituliflorum*) . . . . . 10. *Chionanthoida*, p. 43

Petals glabrous (rarely with 2 or 3 hairs at the apex), usually (but not invariably) longer than 1 mm.:

Petals subreniform, ± 0·9 × 1·5 mm., broader than long; upper receptacle (fig. 3/6, p. 14) cupuliform, small, ± 1·2–1·5 × 1·2–1·5 mm.; stamens 2-seriate; scales (fig. 1/6, p. 12) of a simple 8-celled type; scandent shrubs or lianes . . . . . 2. *Combretastrum*, p. 19

Petals obovate or subcircular (occasionally almost subreniform) to elliptic or spathulate, usually more than 1 mm. long and nearly always as long as broad or longer; stamens 1- or 2-seriate:

Stamens 1-seriate in insertion (or very nearly so) at or near the margin of the disk:

Inflorescences of short (up to 3 cm. long) usually subcapitate or glomeruliform spikes:

Inflorescences glutinous; branchlets and rhachis not fuscous pubescent   .   3. *Macrostigmatea,* p. 21

Inflorescences not glutinous; branchlets and rhachis fuscous pubescent   .   8. *Fusca,* p. 38

Inflorescences of more elongated spikes (usually at least 5 cm. long); cotyledons (where known) usually with connate petioles:

Petals subcircular (blade sometimes almost subreniform); leaves and fruits often somewhat " metallic " in appearance; scales (fig. 1/10, p. 12) 80–180 μ in diameter, divided by many radial and tangential walls to give 16–40 marginal cells, conspicuous (except when hidden by the indumentum); style not expanded at the apex   .   .   .   .   4. *Metallicum,* p. 23

Petals obovate to spathulate; leaves and fruits not " metallic " in appearance, often glutinous (especially when young); style often slightly to considerably expanded at the apex:

Fruits up to 2·5–3·5 × 2·5–3·5 cm., with stipe 0·5–0·7 cm. long; petals (fig. 3/11, p. 14) obovate to spathulate; leaves often 3–4-verticillate; scales (fig. 2/11, 12, p. 13) 40–75 μ in diameter, usually divided by 8 radial walls alone and some secondary radial walls (but virtually never by tangential walls) to give up to 13 marginal cells   .   .   .   .   5. *Glabripetala,* p. 28

Fruits (4–)5(–8) × (3·5–)5(–7) cm., with stipe up to 2·5 cm. long (elongating as the fruit develops); petals (fig. 3/13) obovate-spathulate to spathulate; leaves usually opposite, sometimes 3-verticillate; scales (fig. 2/13) usually with some tangential as well as radial walls   .   .   .   .   6. *Spathulipetala,* p. 31

Stamens clearly 2-seriate in insertion; scales
conspicuous, moderately to very dense:
  Upper receptacle (fig. 3/20, p. 14) some-
    what constricted below the apex;
    fruit broadly 4-winged, usually with
    dark reddish or sometimes yellowish
    scales . . . . . .   9. *Breviramea*, p. 40
  Upper receptacle (fig. 3/26, 27) not con-
    stricted below the apex; inflorescences
    of rather congested often subcapitate
    spikes (but elongated in *C. exalatum*);
    fruit 4-angled or 4-winged (*C.*
    xanthothyrsum); scales usually appear-
    ing somewhat impressed (but
    not in *C. exalatum*) . . .   10. *Chionanthoida*, p. 43

Sect. 1. **Hypocrateropsis** *Engl. & Diels* in E.M. 3: 11 (1899); Stace in J.L.S.
      62: 135 (1969)

Flowers 4-merous. Upper receptacle almost flat, little developed. Petals
linear-elliptic, glabrous. Stamens 8, 1-seriate, inserted at the margin of the
disk. Fruit 4-winged. Cotyledons 2, borne above soil-level (*C. celastroides*,
C. padoides) or arising below soil-level (*C. imberbe*, needing confirmation).
Scales large, usually at least 100 μ in diameter, some rarely only 50 μ, mostly
over 150 μ, roughly circular in outline and divided by many radial and
tangential walls.

Disk glabrous; scales usually not contiguous,
    silvery, golden or rufous . . . .   1. *C. celastroides*
Disk hairy:
  Scales on lower surface of leaf not contiguous,
    golden or rufous:
      Petioles slender, 6–25 mm. long . . .   2. *C. padoides*
      Petioles stout, 2–5 mm. long . . .   3. *C. contractum*
  Scales on both surfaces of leaf very dense,
    contiguous, silvery . . . . .   4. *C. imberbe*

1. **C. celastroides** *Laws.* in F.T.A. 2: 422 (1871); Engl. & Diels in E.M. 3:
12 (1899); F.F.N.R.: 284 (1962); Exell in Kirkia 7: 167 (1970). Type:
Angola, Huila, *Welwitsch* 4370 (LISU, lecto., BM, COI, K!, P, isolecto.)

Small tree 9–12(–20) m. tall, shrub or scandent. Leaves opposite or
subopposite (rarely alternate or pseudo-verticillate); lamina subcoriaceous to
papyraceous, narrowly to broadly elliptic or oblong-elliptic or obovate,
2·5–14 cm. long, 1–8 cm. wide, usually shortly and rather bluntly acuminate
at the apex and cuneate to rounded at the base, glabrous to fairly densely
pubescent especially on the nerves beneath and often with tufts of hairs in
the axils of the lateral nerves beneath, reddish or golden or silvery lepidote
beneath, the scales usually not contiguous; lateral nerves 4–10 pairs; petiole
2·5–20 mm. long. Inflorescences 5–15 cm. long, of terminal panicles of
spikes, with lateral spikes, often unbranched, in the axils of the upper leaves.
Flowers (fig. 3/1, p. 14) yellow or cream or whitish, sweetly scented. Lower
and upper receptacles rufous lepidote, otherwise glabrous. Sepals ovate-
triangular. Petals linear-elliptic, 1·5–2·5 mm. long, 0·3 mm. wide, glabrous.
Stamen-filaments 1·5–2·5 mm. long; anthers 0·4–0·8 mm. long. Disk 2–5·5
mm. in diameter, glabrous, with notches where the filaments are inserted,
without a distinct free margin. Style 1·5–2·8 mm. long. Fruit (fig. 4/1, p. 15)

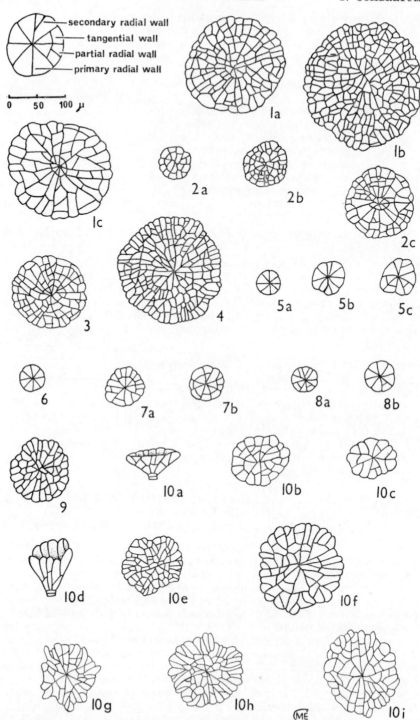

FIG. 1. Scales of *Combretum* species numbered as in text—**1**, *C. celastroides* (**a**, subsp. *celastroides*; **b**, subsp. *orientale*; **c**, subsp. *laxiflorum*); **2 a–c**, *C. padoides*; **3**, *C. contractum*; **4**, *C. imberbe*; **5 a–c**, *C. tanaense*; **6**, *C. umbricola*; **7 a, b**, *C. schumannii*; **8 a, b**, *C. gillettianum*; **9**, *C. tenuipetiolatum*; **10**, *C. collinum* (**a–d**, subsp, *hypopilinum*; **e**, subsp. *elgonense*; **f**, subsp. *binderanum*; **g**, subsp. *gazense*; **h**, subsp. *taborense*; **i**, subsp. *suluense*). Drawn by Mrs. M. E. Church after Stace.

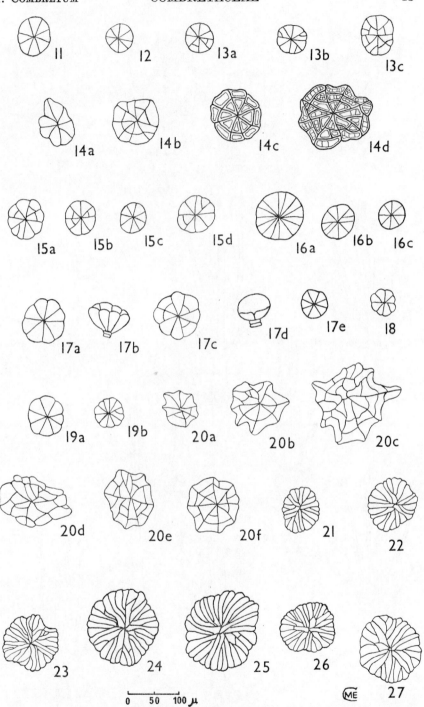

FIG. 2.   Scales of *Combretum* species numbered as in text—11, *C. fragrans*; 12, *C. schweinfurthii*; 13 a–c, *C. zeyheri*; 14 a–d, *C. molle* (thickening of walls shown in c and d); 15, *C. apiculatum* (a–c, subsp. *apiculatum*; d, subsp. *leutweinii*); 16 a–c, *C. acutifolium*; 17, *C. psidioides* (a–d, subsp. *psidioides*; e, subsp. *psilophyllum*); 18, *C. fuscum*; 19, a, b, *C. coriifolium*; 20, *C. hereroense* (a, b, var. *hereroense*; c, var. *villosissimum*; d, subsp. *grotei*; e, subsp. *volkensii* var. *volkensii*; f, subsp. *volkensii* var. *parvifolium*); 21, *C. illairii*; 22, *C. butyrosum*; 23, *C. chionanthoides*; 24, *C, capituliflorum*; 25, *C. xanthothyrsum*; 26, *C. pisoniiflorum*; 27, *C. exalatum*. Drawn by Mrs. M. E. Church after Stace.

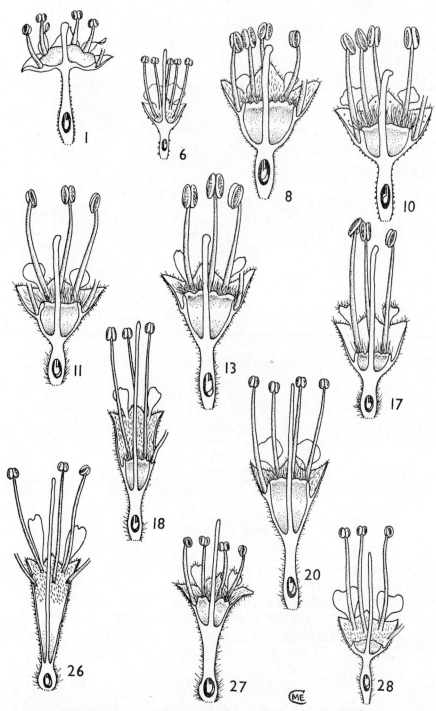

FIG. 3. Longitudinal sections, × 6, of flowers of *Combretum* species numbered as in text—1, *C. celastroides*; 6, *C. umbricola*; 8, *C. gillettianum*; 10, *C. collinum*; 11, *C. fragrans*; 13, *C. zeyheri*; 17, *C. psidioides* subsp. *psidioides*; 18, *C. fuscum*; 20, *C. hereroense*; 26, *C. pisoniiflorum*; 27, *C. exalatum*; 28, *C. mucronatum*. Drawn by Mrs. M. E. Church.

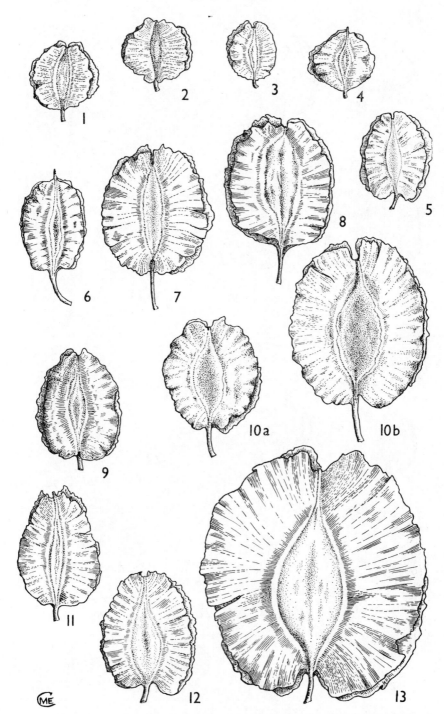

FIG. 4. Fruits, × 1, of *Combretum* species numbered as in text—1, *C. celastroides*; 2, *C. padoides*; 3, *C. contractum*; 4, *C. imberbe*; 5, *C. tanaense*; 6, *C. umbricola*; 7, *C. schumannii*; 8, *C. gillettianum*; 9, *C. tenuipetiolatum*; 10, *C. collinum* (a, subsp. *binderanum*; b, subsp. *suluense*); 11, *C. fragrans*; 12, *C. schweinfurthii*; 13, *C. zeyheri*. Drawn by Mrs. M. E. Church.

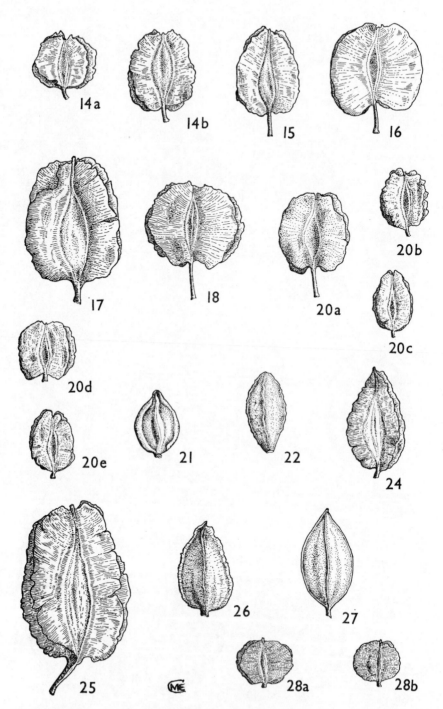

FIG. 5. Fruits, × 1, of *Combretum* species numbered as in text—14a, b, *C. molle*; 15, *C. apiculatum*; 16, *C. acutifolium*; 17, *C. psidioides*; 18, *C. fuscum*; 20, *C. hereroense* (a, var. *hereroense*; b, c, subsp. *grotei*; d, e, subsp. *volkensii* var. *volkensii*); 21, *C. illairii*; 22, *C. butyrosum*; 24, *C. capituliflorum*; 25, *C. xanthothyrsum*; 26, *C. pisoniiflorum*; 27, *C. exalatum*; 28 a, b, *C. mucronatum*. Drawn by Mrs. M. E. Church.

subcircular to transversely broadly elliptic in outline, 1·5–2·5 cm. in diameter, rufous or golden lepidote, otherwise glabrous or nearly so; apical peg 0·5–2·5 mm. long; wings 6–10 mm. broad; stipe 1·5–3 mm. long. Cotyledons 2, with petioles 5–6 mm. long, borne above soil-level. Scales 150–300 μ in diameter (fig. 1/1, p. 12).

subsp. **celastroides**; Exell in Kirkia 7: 168 (1970)

Usually a shrub, sometimes climbing, occasionally a liane, rarely a small tree. Leaves papyraceous or more rarely chartaceous, pubescent especially on the nerves beneath; lamina ± 5–10 cm. long, 1–4·5 cm. wide; scales (fig. 1/1a, p. 12) usually golden or silvery. Disk usually 2·5–3·5(–4) mm. in diameter.

TANGANYIKA. Dodoma District: W. of Dodoma, 5 Mar. 1966, *McCusker* 55!; Mpwapwa District: 112 km. on Iringa–Dodoma road, Feb. 1960, *Procter* 1637!
DISTR. T5; Zambia, Rhodesia, Mozambique, Angola and South West Africa
HAB. Deciduous thicket; 1500 m.

SYN. *C. patelliforme* Engl. & Diels in E.M. 3: 12 (1899), pro parte, excl. *Schlechter* 11957 et t. 1/C. Type: Angola, Huila, *Antunes* A. 155 (B, lecto. †)

subsp. **orientale** *Exell* in Bol. Soc. Brot., sér. 2, 42: 16 (1968) & in Kirkia 7: 169 (1970). Type: Mozambique, Delagoa Bay, *Schlechter* 11957 (BM, holo., COI, K!, iso.)

Small tree or multistemmed shrub. Leaf-lamina up to 6·5 cm. long, 2·5 cm. wide, papery, usually drying olive-green, lightly pubescent beneath with tufts of hairs in the axils of the lateral nerves, sparsely rufous lepidote beneath. Disk 3 mm. in diameter.

TANGANYIKA. Kondoa–Dodoma, 1 Mar. 1928, *B. D. Burtt* 1392!; Dodoma District: Kazikazi, 23 July 1926, *B. D. Burtt* 545! & Manyoni, 17 Feb. 1957, *F. G. Smith* 1386!
DISTR. T4, 5, 7, 8; Zambia, Mozambique and possibly the Transvaal
HAB. Deciduous thicket (one of the principal constituents of the " Itigi Thicket "); 500–1700 m.

SYN. [*C. patelliforme* sensu Engl. & Diels in E.M. 3: 12, t. 1/C (1899), pro parte, quoad specim. *Schlechter* 11957, non Engl. & Diels sensu stricto]
C. trothae Engl. & Diels in E.M. 3: 13, t. 2/A (1899); T.T.C.L.: 138 (1949). Type: Tanganyika, Dodoma/Mpwapwa Districts, Kilimatinde–Mpwapwa, *von Trotha* 175 (B, holo. †, BM, K, fragments of holo.!)

NOTE. (on species as a whole). Almost all the material from East Africa is readily referable to subsp. *orientale*, and the two gatherings referred here to subsp. *celastroides* are not entirely typical, having flowers near the lower part of the range of that subspecies. One gathering of subsp. *orientale*, from Ufipa District, Kawa R., *Richards* 10893!, approaches subsp. *laxiflorum* (Laws.) Exell, a subspecies common in neighbouring parts of Zambia and characterized by a disk ± 4–5·5 mm. across, larger often coriaceous nearly glabrous leaves with a petiole ± 8–20 mm. long.

2. **C. padoides** *Engl. & Diels* in E.M. 3: 13, t. 2/B (1899); T.T.C.L.: 138 (1949); K.T.S.: 146 (1961); F.F.N.R.: 284 (1962); Liben in F.C.B., Combr.: 40 (1968); Exell in Kirkia 7: 170 (1970). Type: Mozambique, Tete, near Boroma, *Menyharth* (Z, lecto., *fide* Exell, *loc. cit.*, K, isolecto.!)

Tree or many-stemmed scandent shrub up to 12 m. tall. Leaves opposite or subopposite; lamina subcoriaceous to papyraceous, elliptic to oblong-elliptic, (1·5–)5–11·5 cm. long, (1·1–)1·5–5·5 cm. wide, apex usually ± acuminate, base acute or obtuse, glabrous or very sparsely pubescent, tufts of hairs in the axils of the lateral nerves beneath absent or inconspicuous, sparsely golden lepidote beneath; lateral nerves 4–8 pairs; petiole 6–25 mm. long. Inflorescences 3–10·5 cm. long, solitary or rarely 2–3 in the axils of the upper leaves (the leaves may fall to give the appearance of a branched terminal panicle). Flowers yellowish or white, scented. Lower and upper receptacles rufous lepidote, otherwise glabrous. Sepals broadly triangular. Petals narrowly elliptic, 1·5 mm. long, 0·25 mm. wide, glabrous. Stamen-filaments 3 mm. long; anthers 0·4 mm. long. Disk 1·5–2 mm. in diameter, pilose, indented where the filaments are inserted, without a distinct free margin. Style 2 mm. long. Fruit (fig. 4/2, p. 15) subcircular to transversely elliptic in outline, 1·5–2 cm. in diameter, rufous or golden lepidote, otherwise glabrous; apical peg absent; wings 7–9 mm. broad, sometimes flushed red; stipe 2–3 mm. long. Cotyledons 2, broadly reniform, petiolate, borne above soil-level. Scales (fig. 1/2, p. 12) similar to *C. celastroides* but mature leaves with scales as small as 50 μ in diameter.

KENYA. Masai District: Loitokitok, 20 July 1962, *Ibrahim* 709!; Kwale District: Marenge Forest, Lungalunga–Msambweni road, 18 Aug. 1953, *Drummond & Hemsley* 3880!; Lamu District: Witu, June 1957, *Rawlins* in *E.A.H.* 11266!
TANGANYIKA. Pare District: Gonjamaore, July 1955, *Semsei* 2113!; Lushoto District: Makuyuni area, June 1935, *Koritschoner* 1044!; Morogoro District: Uluguru Mts., 1 Feb. 1933, *Schlieben* 3371!
DISTR. K6, 7; T2, 3, 5–8; Zaire, Zambia, Rhodesia, Malawi, Mozambique and South Africa (Transvaal)
HAB. Riverine, coastal and swamp forests, also deciduous thickets; 0–1500 m.

SYN. *C. tenuipes* Engl. & Diels in E.M. 3: 13, t. 3/B (1899). Type: South Africa, Transvaal, Barberton, Lowes Creek, *Galpin* 885 (Z, holo., K, iso.!)
C. homblei De Wild. in F.R. 13: 196 (1914). Type: Zaire, Katanga, Kapiri valley, *Homblé* 113 (BR, holo., BM, fragment of holo.!)
C. giorgii De Wild. & Exell in J.B. 67: 100 (1929). Type: Zaire, Bas-Katanga, Kisengwa, *De Giorgi* 167 (BR, holo., BM, fragment of holo.!)
C. minutiflorum Exell in J.B. 68: 245 (1930). Type: Tanganyika, Kilosa District, Kipera [Kipela], *Swynnerton* (BM, holo.)

3. **C. contractum** *Engl. & Diels* in E.J. 39: 491 (1907); T.S.K., ed. 2: 33 (1936); T.T.C.L.: 137 (1949); K.T.S.: 143 (1961). Type: Kenya, Kwale District, Taru plains, *Kassner* 529 (B, holo. †, BM, K, iso.!)

Much-branched shrub up to 3 m. tall. Leaves mainly borne on non-spinose lateral shoots, opposite or subopposite; lamina papyraceous, elliptic to ovate-elliptic, 2–4·5 cm. long, 1–2·5 cm. wide, apex very shortly and bluntly acuminate, base cuneate, shortly pilose above, more sparsely beneath, sometimes only on the nerves beneath and slightly tufted in the axils of the lateral nerves, sparsely scattered golden lepidote beneath; lateral nerves 3–5(–7) pairs; petiole 2–5 mm. long, densely rufous and golden lepidote, sparsely hairy. Inflorescence ± 3 cm. long, of 1–2 single spikes in the axils of leaves on spur shoots, spikes often forming a terminal panicle by the suppression of the upper leaves. Flowers white. Lower and upper receptacles rufous to golden lepidote, otherwise glabrous. Sepals triangular. Petals narrowly oblanceolate, 1·5–2 mm. long, 0·3 mm. wide, glabrous. Stamen-filaments 2·5 mm. long; anthers 0·25 mm. long. Disk 1·5–2 mm. in diameter, margin tomentose. Style 2·5 mm. long. Fruit (fig. 4/3, p. 15) oblong in outline, 1·7–2 cm. long, 1·5 cm. wide, golden lepidote, otherwise glabrous or nearly so; apical peg absent or very short; wings 5 mm. broad; stipe ± 2 mm. long. Scales (fig. 1/3, p. 12) similar to *C. celastroides*.

KENYA. Northern Frontier Province: Dandu, 5 May 1952, *Gillett* 13065!; Kwale District: Mackinnon Road, 15 Apr. 1952, *Bally* 8162!; Kilifi District: near Bamba, 12 Dec. 1962, *Dale* 2049!
TANGANYIKA. Lushoto District: Kivingo, 28 Dec. 1929, *Greenway* 1975!
DISTR. K1, 2, 4, 7; T3; Somali Republic (S.)
HAB. Deciduous bushland, often one of the dominant shrubs in *Acacia, Commiphora* associations; 150–1000 m.

SYN. *C. multiflorum* Pampan. in Bull. Soc. Bot. Ital. 1915: 13 (1915). Types: Somali Republic (S.), Hidlile, *Paoli & Stefanini* 665 & Baidoa–Bur Acaba, *Paoli & Stefanini* 1134 (both FI, syn., K, photo.!)

4. **C. imberbe** *Wawra* in Sitz. Akad. Wiss. Wien, Math.-Nat. Klasse 38: 556 (1860); Engl. & Diels in E.M. 3: 14 (1899); T.T.C.L.: 138 (1949); F.F.N.R.: 284 (1962); Exell in Kirkia 7: 171 (1970). Type: Angola, Benguela, *Wawra* 247 (W, holo., BM, iso.!)

Tree to 33 m. tall or a shrub, branches often becoming spiny; new growth densely rufous lepidote. Leaves opposite or subopposite; lamina silvery, papyraceous to subcoriaceous, narrowly elliptic to elliptic-oblong or elliptic-obovate, (2·5–)5–6(–8·5) cm. long, (1–)2–2·5(–3) cm. wide, apex obtuse or rounded and often mucronate, base obtuse to narrowly cuneate, densely silvery lepidote on both surfaces, otherwise glabrous, scales often contiguous or overlapping; lateral nerves 4–7 pairs, scarcely visible; petiole 4–10 mm. long, densely rufous lepidote. Inflorescence up to ± 10 cm. long, spikes often forming a terminal panicle by suppression of the upper leaves, or unbranched lateral spikes up to 5 cm. long. Flowers yellowish. Lower and upper receptacles densely silvery or rufous lepidote, otherwise glabrous or nearly so. Sepals ovate-triangular. Petals obovate to spathulate, 1–1·2 mm. long, 0·4–0·6 mm. wide, glabrous. Stamen-filaments 1·5–2 mm. long; anthers 0·3–0·4 mm. long. Disk 2–2·5 mm. in diameter, margin densely tomentose. Style 2 mm. long, lower part densely glandular. Fruit (fig. 4/4, p. 15) subcircular to broadly ovate in outline, 1·3–1·8 cm. long, 1·3–1·8 cm. wide, fairly densely to densely silvery lepidote, otherwise glabrous or nearly so, apex pointed; apical peg up to 1 mm. long or absent; wings up to 7 mm. broad; stipe 2–3 mm. long. Cotyledons 2, 15–18 mm. long, 15 mm. wide, with petioles 9–11 mm. long, arising at or below soil-level. Scales (fig. 1/4, p. 12) 120–300 μ in diameter, roughly circular, cells very numerous and small (marginal cells ± 40–100), divided by many radial and tangential walls; cells beneath the scales with conspicuous round papillae.

TANGANYIKA. Kilosa District: Mkata plain, 13 Jan. 1934, *Michelmore* 927!; Morogoro District: between Wami R. and Mvomero, Feb. 1951, *Eggeling* 6014!; near Kilwa, Dec. 1900, *Busse* 553!
DISTR. **T**6, 8; Zambia, Malawi, Mozambique, Rhodesia, Angola, Botswana, South West Africa and South Africa (Transvaal)
HAB. Locally common in wooded grassland on alluvial soils and " black cotton " soils; 0–1000 m.

SYN. *Argyrodendron petersii* Klotzsch in Peters, Reise Mossamb., Bot. 1: 101 (1861). Type: Mozambique, Sena, *Peters* (B, holo. †)
    [*Combretum elaeagnoides* sensu Laws. in F.T.A. 2: 426 (1871), pro parte quoad specim. Angol. *Welwitsch, non* Klotzsch]
    C. truncatum Laws. in F.T.A. 2: 427 (1871). Type: Angola, Mossamedes, *Welwitsch* 4372 (LISU, lecto., K!, BM, isolecto.)
    C. primigenum Engl., E.J. 10: 49 (1888). Type: South West Africa, Hereroland, Usakos, *Marloth* 1264 (B, holo. †, K, iso.!)
    C. petersii (Klotzsch) Engl., P.O.A. C: 290 (1895)
    C. imberbe Wawra var. *dielsii* Engl., E.M. 3: 14, t. 2/C (1899); Diels in E.J. 39: 491 (1907). Types: Tanganyika, Uzaramo District, without precise locality, *Stuhlmann* 6752 & Kilosa District, Kimamba, *Brosig* 21 (both B, syn. †)
    C. imberbe Wawra var. *petersii* (Klotzsch) Engl. & Diels in E.M. 3: 14 (1899); Diels in E.J. 39: 491 (1907)
    C. imberbe Wawra var. *truncatum* (Laws.) Burtt Davy, Fl. Pl. & Ferns Transv. 1: 246 (1926)

Sect. 2. **Combretastrum** *Eichl.* in Mart., Fl. Bras. 14 (2): 115 (1867) [incl. *Olivaceae* Engl. & Diels in E.M. 3: 20 (1899)]; Stace in J.L.S. 62: 143 (1969)

Flowers 4-merous, small. Upper receptacle cupuliform to campanulate. Petals subreniform, glabrous. Stamens 8, 2-seriate. Disk very small, often inconspicuous. Fruit 4-winged (in E. African species) or 4-angled. Scales small, sometimes rather inconspicuous, 40–55(–70) μ in diameter, usually of a simple 8-celled type, usually with the cuticular membrane raised off the cell-plate.

Receptacle lepidote and pubescent; leaf-lamina
    broadly ovate, oblong-ovate or obovate; stipe
    of mature fruits 2–2·5 mm. long . . . 5. *C. tanaense*

Receptacle glabrous in Flora area, except for the
    scales; leaf-lamina elliptic to elliptic-oblong;
    stipe of mature fruits 5–8 mm. long   .   .   6. *C. umbricola*

5. **C. tanaense** *J. J. Clark* in K.B. 1911: 263 (1911); K.T.S.: 146 (1961).
Type: Kenya, Tana R., *Battiscombe* 237 (K, holo.!, EA, iso.)

Scandent shrub or liane, the main branches ending in long leafless whip-
like shoots. Leaves opposite or rarely subopposite; lamina coriaceous,
broadly ovate, oblong-ovate or obovate, 3–15 cm. long, 2–7·6 cm. wide,
acuminate to rounded at apex, base rounded and slightly emarginate to
cuneate, shiny and usually glabrous (except for the scales), often drying a
dark reddish brown above, minutely silvery lepidote but otherwise glabrous
beneath; scales impressed; lateral nerves 6–9 pairs; domatia present in some
of the nerve axils; petiole 4–11 mm. long. Inflorescences terminal panicles
of spikes 3–6 cm. long, in the axils of the upper leaves. Flowers yellowish.
Lower receptacle glutinous, shortly pubescent; upper receptacle 1·5 mm.
long, 1·5 mm. wide, inconspicuously to densely lepidote. Sepals very shallowly triangular or little developed. Petals subreniform to circular, 0·8–1 mm.
long, 1·2–1·5 mm. wide, glabrous. Stamen-filaments 4–5 mm. long; anthers
0·4 mm. long. Disk inconspicuous, glabrous. Style 1·5 mm. long. Fruit
(fig. 4/5, p. 15) oblong-ovate to oblong-elliptic in outline, 2·0–2·3 cm. long,
1·8–2·0 cm. wide, apex retuse, base shallowly cordate, minutely lepidote,
otherwise glabrous; apical peg inconspicuous; stipe 2–2·5 mm. long. Scales
(fig. 1/5, p. 12) 55–80 μ in diameter, simple 8-celled type with few tangential
walls.

KENYA. Fort Hall District: Thika R., near Mabaloni hill, Jan. 1972, *Gillett* 19433!;
    Embu/Kitui District: Tana R., Apr. 1910, *Battiscombe* 237!
DISTR. **K4**; not known elsewhere
HAB. Riverine forest; 900–1140 m.

NOTE. This species is closely allied to *C. cuspidatum* Benth. from West Africa, with
    which it has been confused, e.g. Stace in J.L.S. 62: 143 (1969), due to the absence of
    fruiting material. More collections are required.

6. **C. umbricola** *Engl.*, P.O.A. C: 288 (1895)* & E.M. 3: 23, t. 5/D (1899);
Burtt Davy, Check-lists Brit. Emp. 1, Uganda: 36 (1935); T.T.C.L.: 138
(1949); Liben in F.C.B., Combr.: 56 (1968); Exell in Kirkia 7: 173 (1970).
Type: Tanganyika, Tanga District, Doda, *Holst* 2965 (B, holo. †, K, iso.!)

Small tree or shrub, or in East Africa a scandent shrub or liane. Leaves
opposite; lamina chartaceous to subcoriaceous, elliptic to elliptic-oblong,
3–14(–16) cm. long, 2–6(–6·5) cm. wide, usually acuminate, base cuneate or
obtuse to rounded, minutely verrucose and usually shiny and glabrous above,
minutely silvery lepidote but otherwise glabrous beneath; lateral nerves
6–10 pairs; petiole up to 9(–13) mm. long. Inflorescence terminal and
axillary panicles of spikes 2–3 cm. long, in the axils of the upper leaves.
Flowers (fig. 3/6, p. 14) white, scented. Lower receptacle glabrous or pubescent; upper receptacle 1·2–1·5 mm. long and wide, not conspicuously lepidote.
Sepals very shallowly triangular or little developed. Petals subreniform,
0·9 mm. long, 1·5 mm. wide, glabrous. Stamen-filaments 3–4 mm. long;
anthers 0·4 mm. long. Disk inconspicuous, glabrous. Style 4–4·5 mm. long,
sometimes somewhat flattened near apex. Fruit (fig. 4/6, p. 15) broadly
elliptic in outline, up to 2·8 cm. long and 2·3 cm. wide, minutely lepidote,
otherwise usually glabrous; wings 7 mm. wide; apical peg 1–2 mm. long;

---

* The name first appears in Abh. Preuss. Akad. Wiss. 1894: 7 (1894), without description.

stipe ± 5–8 mm. long. Scales (fig. 1/6, p. 12) of a simple 8-celled type, 40–50(–70) μ in diameter, frequent in the areoles of the lower leaf-surface.

Uganda. Toro District: Mpanga R., 20 June 1906, *Bagshawe* 1059! & Nyamwamba valley, near Kilembe, 19 Dec. 1934, *G. Taylor* 2519!; Toro/Mubende Districts: Muzizi R., 3 Dec. 1906, *Bagshawe* 1330!
Tanganyika. Tanga District: Mtimbwani, 6 Dec. 1935, *Greenway* 4208! & Ngole, 9 June 1937, *Greenway* 4941!; Pangani District: Bushiri Estate, 4 June 1950, *Faulkner* 582!
Zanzibar. Pemba I., *Barraud*!
Distr. U2, 2/4; T3; P; Zaire, Mozambique
Hab. Riverine forest and coastal *Brachystegia spiciformis* woodland; 10–50 m. and 660–1050 m.

Note. Despite the distance separating the main areas of distribution, the specimens are taxonomically inseparable. It is possible that the Zaire and Uganda populations belong to what Engler (P.O.A. C: 288 (1895)) called *C. olivaceum* Engl. of which the type, *Stuhlmann* 2851 (B, holo.) from Zaire, SW. of Lake Albert, has been destroyed.

Sect. 3. **Macrostigmatea** *Engl.* & *Diels* in E.M. 3: 24 (1899); Stace in J.L.S. 62: 159 (1969)

Flowers 4-merous, glutinous, in subcapitate spikes. Upper receptacle cupuliform to infundibuliform. Petals ovate, subcircular, obtriangular or obovate, glabrous. Stamens 8, 1-seriate, inserted at the margin of the disk. Disk glabrous with only a very short free margin or with a pilose margin free for ± 1 mm. Style sometimes (? always) with an expanded stigma, the latter quickly withering. Cotyledons (known only in *C. kirkii* from the Zambezi valley) 2, borne above soil-level. Scales 35–85 μ in diameter, usually of a simple 8-celled type with the ± constant addition of a number of tangential and extra radial walls giving up to 16 marginal cells.

Disk glabrous; branching not divaricate; fruit up to 3·5 cm. in diameter, subcircular in outline, glutinous when young, stipe up to 1·5 cm. long . . . . . .   7. *C. schumannii*
Disk pilose; branching divaricate; fruit 3–4 cm. long, oval to subcircular in outline, tomentellous when young; wings decurrent at the base into the 1·2–2 cm. long stipe. . .   8. *C. gillettianum*

7. C. schumannii *Engl.* [in Abh. Preuss. Akad. Wiss. 1894: 34 (1894), *nomen.*], P.O.A. C: 289 (1895) & Engl. & Diels in E.M. 3: 24, t. 6/C (1899); T.S.K.: 22 (1926) & ed. 2: 32 (1936); T.T.C.L.: 138 (1949); K.T.S.: 146 (1961); Exell in Kirkia 7: 177 (1970). Type: Tanganyika, Lushoto District, Bwiti [Buiti], *Holst* 2375 (B, holo. †, K, ? iso.! *)

Shrub or tree up to 18 m. high. Bark buff-brown, spongy, scaling off in large longitudinal flakes; wood purplish black. Leaves opposite; lamina papyraceous to chartaceous, elliptic or oblong-elliptic, 6–12(–15) cm. long, 2–4·5(–7·5) cm. wide, apex rather blunt and somewhat acuminate, base cuneate to rounded, verrucose above, otherwise glabrous and shiny, inconspicuously silvery lepidote beneath, sparsely pubescent on midrib and with hairs in axils of lateral nerves, otherwise glabrous; lateral nerves 3–8 pairs, alternate or subopposite; petiole up to 8(–10) mm. long. Inflorescence subglobose-fasciculate axillary spikes 1–1·5 cm. in diameter. Flowers yellow, scented. Lower receptacle glutinous, glabrous; upper receptacle 1·5 mm. long and wide, with 4 pouch-like swellings alternating with the position of

* The Kew sheet gives the locality as Zimbili [Simbili], which is, however, in the Bwiti area.

the petals, not conspicuously lepidote. Sepals shallowly triangular, little developed. Petals suborbicular, 1·3 mm. long, 1·5 mm. wide, glabrous. Stamen-filaments 4 mm. long; anthers 0·5 mm. long. Disk glabrous, margin scarcely produced. Style 5·5 mm. long, expanded near apex. Fruit (fig. 4/7, p. 15) yellowish green, subcircular in outline, up to 3·5 cm. in diameter, glutinous when young; stipe 1–2 cm. long. Cotyledons 2, borne above ground, sessile, obreniform, shortly and broadly acuminate. Scales (fig. 1/7, p. 12) 45–75 μ in diameter.

KENYA. Kitui District: near Mombasa, 15 Apr. 1958, *Trapnell* 2397!; Kwale District: Mrima Hill, 7 Sept. 1957, *Verdcourt* 1910!; Coast Province, without precise locality, *Webber* 435!
TANGANYIKA. Lushoto District: Kwamkuyu Falls, 23 July 1953, *Drummond & Hemsley* 3423! & Kwalukonge, *Gillman* 949!; Tanga District: Ngomeni, Sept. 1955, *Semsei* 2282!
DISTR. K4, 7; T2, 3, 5, 6, 8; Malawi, Mozambique
HAB. A wide range of habitats from lowland rain-forest, riverine forest and *Brachystegia* woodland to valley grassland; 0–1150 m.

SYN. *C. macrostigmateum* Engl. & Diels in E.M. 3: 24 (1899); T.T.C.L.: 138 (1949). Type: Tanganyika, Dodoma District, Saranda, *Fischer* 291 (B, holo. †)

NOTE. As more material becomes available it is becoming difficult to maintain *C. engleri* Schinz (*C. parvifolium* Dinter *non* Engl., *C. myrtillifolium* Engl. and *C. chlorocarpum* Exell) from Angola, South West Africa, Botswana and Zambia as a distinct species from *C. schumannii* Engl. The differences which exist (smaller leaves and fruits) are those to be expected under drier climatic conditions and it seems probable that *C. engleri* will have to be reduced to subspecific or varietal status.

8. **C. gillettianum** *Liben* in B.J.B.B. 35: 177 (1965) & in F.C.B., Combr.: 70 (1968); Exell in Kirkia 7: 178 (1970). Type: Zambia, Mpulungu, *Glover in Bredo* 6143 (BR, holo.)

Straggling shrub or small tree up to 4 m.; branches nearly at right-angles, glabrous. Leaves opposite; lamina papyraceous to chartaceous or sub-coriaceous, narrowly elliptic, 2·5–5 cm. long, 1–2 cm. wide, acute or obtuse, rounded or emarginate at the base, glabrous except for some indumentum on the nerves when young and tufts of hairs in the axils of the lateral nerves beneath, conspicuously lepidote when young but inconspicuously so when mature; lateral nerves 3–6 pairs; petiole 1–2(–3·5) mm. long, puberulous. Inflorescence of short, 1 cm. long, subcapitate spikes in the axils of the young leaves or forming a short panicle by suppression of the upper leaves. Flowers (fig. 3/8, p. 14) yellowish or yellow-green. Lower receptacle 2–2·5 mm. long, tomentellous; upper receptacle cupuliform, 3·5–4(–4·5) mm. long, 3–4 mm. wide, appressed pubescent, often with 4 tomentellous nerves running up into the sepals. Sepals triangular. Petals obtriangular to obovate or sub-circular, 1·3 mm. long, 1 mm. wide, glabrous. Stamen-filaments 3·5–4 mm. long; anthers 1–1·2 mm. long. Disk with a pilose margin free for ± 1 mm. Style 1·5–2 mm. long, somewhat swollen at the apex and apparently con-siderably expanded for a short period. Fruit (fig. 4/8, p. 15) oval to sub-orbicular in outline, 3–4 cm. long, 2·5–3 cm. wide, tomentellous when young, not conspicuously lepidote, apex somewhat emarginate; apical peg short, up to 0·6 mm.; wings 9–12 mm. wide, decurrent at the base in the 1·2–2 cm. long slender stipe. Scales (fig. 1/8, p. 12) 35–60(–70) μ in diameter, with 8–12 marginal cells.

TANGANYIKA. Ufipa District: Kasanga, old German road to Mbala [Abercorn], 19 June 1957, *Richards* 10166! & Sumbawanga, 24 Nov. 1959, *Richards* 11813!
DISTR. T4; Zaire, Zambia
HAB. *Brachystegia* woodland and deciduous thicket; 900 m.

SYN.   *C. sp.* sensu Hutch., Botanist in S. Afr.: 520 (1946)
      [*C. engleri* sensu F. White, F.F.N.R.: 285 (1962), pro parte quoad specim.
      Angus 773A & *White* 3679, *non* Schinz]

Sect. 4. **Metallicum** *Exell & Stace* in Bol. Soc. Brot., sér. 2, 42 : 22 (1968);
                  Stace in J.L.S. 62 : 145 (1969)

SYN. Sect. *Glabripetala* Engl. & Diels in E.M. 3 : 43 (1899), pro parte

Flowers 4-merous.   Upper receptacle campanulate at the base and cupuliform at the apex, clearly divided into 2 regions.   Petals transversely elliptic
to obovate or subcircular, usually glabrous, rarely with a few hairs at the apex.
Stamens 8, 1-seriate, inserted at the margin of the disk.   Disk with a free
pilose margin.   Fruit 4-winged, brown or greyish brown or reddish grey to
dark purple, lepidote, glabrous (apart from the scales) to densely hairy,
usually somewhat " metallic " in appearance.   Cotyledons 2, borne below
soil-level on a stalk formed by the connate petioles or above soil-level with
short free petioles (*C. collinum* subsp. *dumetorum* from Angola and Zambia).
Scales 55–180 μ in diameter, usually extensively divided by many radial and
tangential walls to give 16–40 marginal cells.   Scales discoid except in *C.
collinum* subspp. *dumetorum* and *hypopilinum* where they are funnel-shaped,
thus appearing smaller, and also divided with 10–16 marginal cells.

Leaf-blades elliptic, up to 8 cm. long, papery,
        hairless, pustulate, with only silvery scales
        beneath; petioles (in dried specimens) 0·5–
        0·8 mm. in diameter; fruits ± 3 cm. long,
        ovate, the body hoary, with few scattered red
        scales, wings copper-brown .   .   .   .   9. *C. tenuipetiolatum*
Leaf-blades variable but usually larger and thicker
        textured, always with a conspicuous covering
        of hairs or scales beneath; petioles generally
        thicker; fruits 2·5–5·9 cm. long, variable in
        shape, indumentum, colour, etc. .   .   .   10. *C. collinum*

9. **C. tenuipetiolatum** *Wickens* in K.B. 25 : 182, fig. 2 (1971).   Type:
Tanganyika, Tanga District, Potwe, *Semsei* 3142 (K, holo.!, EA, TFD, iso.)

Small tree; branches mouse-grey; branchlets reddish brown, lightly
pubescent.   Leaves opposite; lamina elliptic, up to 8 cm. long, 3·5 cm. wide,
apex shortly obtuse-acuminate, base cuneate, pustulate, glabrous except for
the sparse scales and pubescent domatia in the nerve-axils; midrib prominent
beneath; lateral nerves 5–6 pairs; petiole up to 9 mm. long, slender.   Inflorescences solitary axillary spikes 5–10 cm. long.   Flowers yellow.   Lower
receptacle 3 mm. long, lightly pubescent to moderately densely rufous
pubescent; upper receptacle 5 mm. long, infundibuliform below for 3 mm.,
then widening rather abruptly to a 3 mm. broad campanulate crown,
sparsely pubescent.   Sepals deltate, 0·7 mm. long.   Petals obtriangular,
0·8 mm. long, 1 mm. wide, emarginate.   Stamen-filaments 4 mm. long;
anthers 0·9 mm. long.   Disk with pilose margin free for 1 mm.   Fruit (fig.
4/9, p. 15) ovate in outline, 3 cm. long, 2·5 cm. wide; body hoary with a few
scattered red scales; wings firm and copper-brown; apical peg absent; stipe
± 10 mm. long, slender.   Scales (fig. 1/9, p. 12) ± 120 μ in diameter.

KENYA.   Mombasa District: without precise locality, *Wakefield*!; Kilifi District: Rabai,
      Joanna in E.A. Nat. Hist. Soc. 5917!
TANGANYIKA.   Tanga District: Potwe Forest, 30 Dec. 1960, *Semsei* 3142!
DISTR.   **K7**; **T3**; not known elsewhere

HAB. Apparently coastal forest, seemingly rather rare; below 150 m.

NOTE. Further gatherings of this interesting species are required.

10. **C. collinum** *Fresen.* in Mus. Senckenb. 2: 153 (1837); Engl., P.O.A. C: 290 (1895); Engl. & Diels in E.M. 3: 56 (1899); Okafor in Bol. Soc. Brot., sér. 2, 41: 137–150 (1967); Exell in Kirkia 7: 179 (1970). Type: Ethiopia, Begemeder, near Gondar, Kulla, *Rueppell* (FR, holo.)

Small tree up to ± 12 m. high or a coppicing shrub; bark reddish brown or pale yellow. Leaves opposite or alternate or verticillate; lamina coriaceous or subcoriaceous, very variable in shape, up to 22 cm. long and 8 cm. wide, upper surface drying reddish olive or yellowish brown, usually somewhat " metallic " in appearance, lower surface green, buff or silvery, glabrous (except for scales) to densely tomentose, conspicuously lepidote except when the scales are hidden by the indumentum; scales contiguous or not; lateral nerves 6–20 pairs, reticulation sometimes prominent; petiole 1–4 cm. long. Inflorescences simple spikes or panicles up to 10 cm. long, axillary or supra-axillary from bracts or reduced leaves on the current year's shoots (the panicle appears to be derived by the suppression of these leaves or bracts), glabrous or pubescent. Flowers (fig. 3/10, p. 14) yellow, cream or white, fragrant. Lower receptacle glabrous (except for scales) or hairy, lepidote; upper receptacle campanulate at the base and cupuliform at the apex, generally 3·5 mm. long, 2·5 mm. wide, or, in subsp. *gazense*, 5–5·5 mm. long, 3·5–4 mm. wide, glabrous (except for scales) to tomentose, lepidote. Sepals broadly triangular. Petals transversely elliptic to obovate or subcircular, 1·5–2·5 mm. long, 1–2·5 mm. wide, somewhat emarginate, with a claw up to 1 mm. long, glabrous (very rarely with a few hairs at the apex). Stamen-filaments 4–4·5(–6·5 in subsp. *gazense*) mm. long; anthers 0·9 mm. long. Disk 2 mm. in diameter, with pilose margin free for ± 1 mm. Style up to 7·5 mm. Fruit (fig. 4/10, p. 15) brown, reddish brown or greyish brown to dark purple, usually somewhat " metallic " in appearance, variable in shape, 2·5–5·9 cm. long, lepidote with the scales often contiguous, glabrous (except for scales) to densely pubescent, glossy or dull; wings variable in breadth; stipe up to 8(–15) mm. long. Cotyledons 2, arising below soil-level on a stalk formed by their connate petioles, free part of petiole 2 mm. long (subsp. *dumetorum* from Angola differs in having a " normal " type of germination with the cotyledons arising above soil-level on short, quite free petioles and may be specifically distinct). Scales (fig. 1/10, p. 12) discoid or funnel-shaped, 55–180 μ in diameter, usually extensively divided by many radial and tangential walls to give ± 10–40 marginal cells.

DISTR. (of species as a whole). **U**1–4; **K**2–5; **T**1–8; widespread in tropical and sub-tropical Africa

NOTE. The 54 " species " contained in this aggregate have been separated by Okafor (Bol. Soc. Brot., sér. 2, 41: 137–150 (1967)) into 11 reasonably distinct subspecies, 6 of which occur in East Africa. Although his treatment has been followed, it may well be that further study will provide a more satisfactory method for dealing with this very complex aggregate.

KEY TO INFRASPECIFIC VARIANTS

Leaves in whorls of 3 or 4; fruit usually less than
    3·6 cm. long:
Leaves distinctly hairy beneath, at least on
        the midrib; scales, if visible, not contiguous;
        fruit villous or velutinous, lepidote:
Scales (fig. 1/10a, d, p. 12) on lower leaf-
            surface cup- or bowl-shaped; leaf-vena-

tion very prominent; fruit velutinous,
with widely scattered red scales . . a. subsp. **hypopilinum**
Scales on lower leaf-surface discoid; fruit
villous, often with a dense covering of red
scales in the centre . . . . b. subsp. **elgonense**
Leaves glabrous beneath or almost so (except
for scales); scales contiguous or almost so;
fruit not conspicuously pubescent, densely
covered with red scales in the centre . . c. subsp. **binderanum**
Leaves all opposite or alternate; fruit more than
3·5 cm. long:
Leaves with prominent reticulation and long
spreading hairs; inflorescence branched
(in East Africa) or simple; flowers large,
tomentose; fruit glabrous except for dense
cover of red scales . . . . . d. subsp. **gazense**
Leaves glabrous (except for scales) or almost so;
inflorescence simple or branched:
Lower leaf-surface silvery with contiguous
silvery scales and reddish venation;
inflorescence branched; fruit with con-
spicuous red scales, otherwise glabrous . e. subsp. **taborense**
Lower leaf-surfaces never silvery, scales not
contiguous; inflorescence simple; fruit
densely lepidote, otherwise glabrous to
slightly pubescent . . . . f. subsp. **suluense***

a. subsp. **hypopilinum** (*Diels*) *Okafor* in Bol. Soc. Brot., sér. 2, 41: 142 (1967). Types:
Central African Republic, Ndellé, *Chevalier* 7431, 8462 & 8466 (P, syn.) & 7469 (P, syn.,
K, isosyn.!) & Chad, Fort Archambault, *Chevalier* 1047 (P, syn.)

Young branches angular with yellow-brown pubescence. Leaf-lamina coriaceous,
variable in shape from narrowly elliptic to oblong, up to 22 cm. long and 6(–8) cm. wide,
apex acute to retuse, base cuneate, upper surface drying greenish brown, lower surface
yellow-brown, shortly pubescent to tomentose beneath, scales (fig. 1/10a–d, p. 12)
cup-shaped, scattered, rarely contiguous and frequently difficult to see; lateral nerves
12–20 pairs, reticulation very prominent beneath on mature leaves; petiole 1·5 cm. long,
pubescent. Inflorescence simple. Lower and upper receptacles tomentose. Fruit
reddish grey or brown, distinctly rusty velutinous, with a few widely scattered red
scales.

UGANDA. W. Nile District: near Chei Hill, Sept. 1937, *Eggeling* 3411! & Leo, Koich R.,
Sept. 1937, *Eggeling* 3414!
DISTR. **U**1; Guinée to Central African Republic and Sudan
HAB. Wooded grassland, not common; ± 1000 m.

SYN. *C. verticillatum* Engl., E.M. 3: 52, t. 16/B/a–h (1899); F.P.S. 1: 203 (1950);
I.T.U., ed. 2: 89 (1952); Liben in F.C.B., Combr.: 73 (1968). Types: Sudan,
Bahr el Ghazal, Jur [Gir], *Schweinfurth* 1411 (B, syn. †, K, isosyn.!) & Jur
[Seriba Ghattas], *Schweinfurth* 1595 (B, syn. †, K, isosyn.!)
C. hypopilinum Diels in E.J. 39: 947 (1907)
C. kottoense Exell in Bull. Soc. Linn. Normand. 1936, sér. 8, 9: 133 (1937). Type:
Central African Republic, Haute-Kotto, *Le Testu* 3675 (BM, holo.!, P, iso.)
C. flaviflorum Exell in Bull. Soc. Linn. Normand. 1936, sér. 8, 9: 133 (1937),
pro parte quoad *Tisserant* 961 tantum. Type: Central African Republic,
Haute-Kotto, 40 km. N. of Moroubas, *Tisserant* 961 (P, holo., BM, iso.!)

b. subsp. **elgonense** (*Exell*) *Okafor* in Bol. Soc. Brot., sér. 2, 41: 142 (1967). Type:
Kenya, Elgon, *Lugard* 524 (K, holo.!, BM, EA, iso.!)

Young branches terete or subangular, lightly tomentose, the hairs white or cream.
Leaf-lamina coriaceous, elliptic to broadly elliptic, up to 11(–17) cm. long and 6(–8) cm.

* In the absence of fruits subsp. *suluense* is easily confused with subsp. *binderanum* if
mature well-developed leaves are not present.

wide, apex obtuse to acute, base cuneate to rounded; upper surface drying reddish brown to purplish brown, lower surface olive-green, white to yellowish villous beneath; scales (fig. 1/10e, p. 12) discoid, yellowish, not contiguous; lateral nerves 7–10, prominent beneath, visible above; petiole 1–1·5 cm. long, villous. Inflorescence simple. Lower and upper receptacles densely pubescent to tomentose. Fruit dark red or purple, often with a dense cover of red scales in the centre, but sometimes without, villous.

UGANDA. Teso District: Serere, Jan. 1932, *Chandler* 638!; Mbale District: Busaba, July 1926, *Maitland* 1135!
KENYA. Trans-Nzoia District: Kitale, Mar. 1953, *Brockington* 46! & Kitale, base of Mt. Elgon, 17 Apr. 1943, *Bally* 2481!; N. Kavirondo District: Malikisi [Malakisi], Feb. 1956, *Templer* 66!
TANGANYIKA. Mwanza District: Geita, 19 June 1937, *B. D. Burtt* 6568! & Usagara, 19 Feb. 1953, *Tanner* 1218!; Buha District: Mkalinzi, Nov. 1956, *Procter* 566!
DISTR. Ul, 3; K2, 3, 5; Tl, 4, 7; Sudan, Zaire, Zambia
HAB. Common in wooded grassland on a wide range of soils; 1100–2300 m.

SYN. *C. laboniense* M. B. Moss in K.B. 1929: 195 (1929); F.P.S. 1: 202 (1950). Type: Sudan, Equatoria, Imatong Mts., near Laboni Forest, *Chipp* 45 (K, holo.!, BM, iso.)
C. elgonense Exell in K.B. 1932: 491 (1932); K.T.S.: 144 (1961)
C. sp. near C. elgonense sensu Burtt Davy, Check-lists Brit. Emp. 1, Uganda: 36 (1935)
C. kabadense Exell in J.B. 75: 165 (1937); F.P.S. 1: 202 (1950). Type: Sudan, Equatoria, Mongalla, Kabada Hills, *Dandy* 460 (BM, holo.!)
C. abercornense Exell in J.B. 77: 167 (1939). Type: Zambia, Mbala [Abercorn] District, Kalambo Falls, *B. D. Burtt* 5963 (BM, holo.!, EA, K!, iso.)
C. mwanzense Exell in J.B. 77: 168 (1939); T.T.C.L.: 141 (1949). Type: Tanganyika, Uzinza area W. of Mwanza, *B. D. Burtt* 6455 (BM, holo.!, EA, K!, iso.)
[*C. mechowianum* sensu K.T.S.: 145 (1961); F.F.N.R.: 286 (1962), pro parte, *non* O. Hoffm.]

c. subsp. **binderanum** (*Kotschy*) *Okafor* in Bol. Soc. Brot., sér. 2, 41: 141 (1967). Type: Sudan, Bahr el Ghazal, W. of Shambe [Gaba Schambil], *Binder* 46 (W, holo., BM, fragment!)

Young branches angular or terete. Leaf-lamina coriaceous, variable in shape from narrowly elliptic to obovate, up to 13(–16) cm. long and 6(–8) cm. wide, apex usually acute and base cuneate, upper surface drying dark to light olive-brown, lower surface light yellow-brown, usually glabrous on both young and old leaves but occasionally sparsely pubescent on the midrib; scales (fig. 1/10f, p. 12) peltate, contiguous or almost so; lateral nerves 7–8 pairs, reticulation usually not prominent on mature leaves in East Africa; petiole 2·5 cm. long, glabrous. Inflorescence simple. Lower receptacle with dense mass of red scales, but scales only scattered on upper receptacle; receptacles sometimes sparsely pubescent in East Africa. Fruit (fig. 4/10a, p. 15) reddish brown, with red scales in the centre, not conspicuously pubescent.

UGANDA. Lango District: near Lira, Oct. 1930, *Brasnett* 37!; Bunyoro District: near Masindi, 11 Aug. 1955, *Langdale-Brown* 1414!; Busoga District: Namaira, 13 June 1926, *Maitland*!
KENYA. W. Suk District: Kongelai, June 1961, *Lucas* 156!; Trans-Nzoia District: Kitale–Kapenguria, 3 Feb. 1947, *Bogdan* 264!; Central Kavirondo District: Ugenya Location, 15 Aug. 1944, *Davidson* 207 in *Bally* 4414!
TANGANYIKA. Mwanza District: Ukerewe I., *Conrads* 984!; Tabora District: Kasisi valley, *Wigg*!; Mpanda District: Kapapa Camp, 28 Oct. 1959, *Richards* 11606!
DISTR. Ul–4; K2–5; Tl, 4–7; Ivory Coast to N. Nigeria, Sudan and Ethiopia
HAB. Common in wooded grassland on a wide range of soils; 50–2200 m.

SYN. *C. elaeagnifolium* Oliv. in App. to Speke, Journ. Disc. Source Nile: 634 (1863), nom. nud., based on *Grant* from Madi, Equatoria Province, Sudan (K!)
C. binderanum Kotschy in Sitz. Akad. Wiss. Wien, Math.-Nat. Klasse 51: 363, t. 5 (1865), as " *binderianum* "; Burtt Davy, Check-lists Brit. Emp. 1, Uganda: 35 (1935); T.T.C.L.: 138 (1949); F.P.S. 1: 199 (1950); I.T.U., ed. 2: 86 (1952), pro parte; K.T.S.: 143 (1961); Liben in F.C.B., Combr.: 78 (1968)
C. populifolium Engl. & Diels in E.M. 3: 54 (1899); T.S.K., ed. 2: 32 (1936). Types: Sudan, Bahr el Ghazal, Jur [Seriba Ghattas], *Schweinfurth* 1374 (B, syn. †, K, isosyn.!) & Tanganyika, Singida/Dodoma District, Turu, Ulagesa, *von Trotha* 148 (B, syn. †)
C. karaguense Engl. & Diels in E.M. 3: 55, t. 18/A/a–c (1899); Burtt Davy, Check-lists Brit. Emp. 1, Uganda: 36 (1935). Type: Tanganyika, Bukoba District, Karagwe, *Scott Elliott* 8130 (B, holo. †, BM, K!, iso.)

? C. frommii Engl., V.E. 3 (2): 700 (1921); T.T.C.L.: 140 (1949).   Type: Tanganyika, mountains SE. of Lake Tanganyika, *Muenzner* 269 (B, holo. †)

d. subsp. **gazense** (*Swynn. & Bak. f.*) *Okafor* in Bol. Soc. Brot., sér. 2, 41: 145 (1967). Type: Mozambique, Manica e Sofala, Mt. Umtereni, *Swynnerton* 587 (BM, holo. !, K, iso. !)

Young branches terete, densely olive-brown velutinous.   Leaf-lamina chartaceous to coriaceous, variable in shape from narrowly elliptic to ovate or obovate, up to 12 cm. long, 5·5 cm. wide, apex acute or obtuse, base cuneate or rounded; upper surface drying reddish or yellowish brown, lower surface buff, tomentose beneath, indumentum frequently concealing the scales; lateral nerves 6–9 pairs, reticulation frequently close and prominent, but not usually so in East Africa; petiole 0·7–1(–1·5) cm. long, pubescent. Inflorescence branched (in East Africa) or simple.   Flowers larger than for the other subspecies, upper receptacle 5–5·5 mm. long; receptacles tomentose, the hairs obscuring the red scales of the lower receptacle.   Fruit glabrous except for dense cover of red scales.

TANGANYIKA.   Singida District: Gandajega R., Sept. 1935, *B. D. Burtt* 5251 ! & Singida, 20 Feb. 1934, *Maclean* in *B. D. Burtt* 5071 !; Iringa District: Great Ruaha R. 12 km. W. of Kidatu bridge, 3 Sept. 1970, *Thulin & Mhoro* 827 !
DISTR.   T5, 7; Zaire (Katanga), Zambia, Rhodesia, Malawi, Mozambique, South Africa (Transvaal) and South West Africa
HAB.   Wooded grassland and deciduous bushland; 450–1700 m.

SYN.   *C. bajonense* Sim, For. Fl. Port. E. Afr.: 63, t. 63/B–C (1909).   Type: Mozambique, Maganja da Costa, *Sim* 5715 (PRE, iso., renumbered No. 20901)
  C. gazense Swynn. & Bak. f. in J.L.S. 40: 68 (1911)
  C. album De Wild. in F.R. 13: 195 (1914).   Type: Zaire, Katanga, Lubumbashi [Elisabethville], *Bequaert* 513 (BR, holo.)
  C. ritschardii De Wild. & Exell in De Wild., Pl. Bequaert. 5: 366 (1932).   Type: Zaire, Katanga, Lubumbashi [Kiluba area], *Ritschard* 1478 & 1479 & Lualaba, Ruwa, *Hock* (all BR, syn.)
  C. singidense Exell in J.B. 77: 169 (1939); T.T.C.L.: 141 (1949).   Type: Tanganyika, Singida District, Gandajega R., *B. D. Burtt* 5251 (BM, holo. !, FHO, K !, iso.)
  C. mechowianum O. Hoffm. subsp. *gazense* (Swynn. & Bak. f.) Duvign. in B.S.B.B. 88: 81, t. 8/c–d (1956); Liben in F.C.B., Combr.: 77 (1968)
  [*C. mechowianum* sensu F.F.N.R.: 286 (1962), pro parte, *non* O. Hoffm. sensu stricto]

e. subsp. **taborense** (*Engl.*) *Okafor* in Bol. Soc. Brot., sér. 2, 41: 144 (1967).   Type: Tanganyika, Tabora, *Stuhlmann* 506 (B, holo. †)

Young branches angular, pubescent.   Leaf-lamina coriaceous to chartaceous, narrowly elliptic to ovate, up to 16 cm. long and 7 cm. wide, frequently acuminate, base rounded or cuneate, upper surface drying olive- to reddish brown, lower surface silvery or yellowish grey, usually glabrous (except for scales), sometimes pubescent on the nerves beneath; scales (fig. 1/10h, p. 12) peltate, white, contiguous; lateral nerves 7–10 pairs, venation distinct and darker than undersurface of leaf; petiole 1–1·5(–2·5) cm. long, glabrous or lightly pubescent.   Inflorescence branched, pubescent.   Lower and upper receptacles tomentose or sometimes pubescent, lower receptacle occasionally with conspicuous red scales.   Fruit glabrous, red-brown and densely lepidote.

TANGANYIKA.   Mpanda District: Utahya–Kibwesa, 5 Aug. 1958, *Mgaza* 176 !; Chunya District: Lupa Forest Reserve, 2 Aug. 1962, *Boaler* 644 !; Masasi District: Ruvuma, Aug. 1934, *Hubbert* 2027 !
DISTR.   T2, 4–8; Zaire (Katanga), Malawi, N. Mozambique, NE. Zambia & E. Rhodesia
HAB.   *Brachystegia* woodland and wooded grassland; 450–1400 m.

SYN.   *C. taborense* Engl., P.O.A. C: 290 (1895); T.T.C.L.: 141 (1949)
  C. goetzenianum Diels in E.J. 39: 500 (1907); T.T.C.L.: 140 (1949).   Type: Tanganyika, Kilwa District, Liwale R., *Busse* 570 (B, holo. †, BM !, EA, iso.)
  C. psammophilum Engl. & Diels in E.J. 39: 502 (1907).   Type: Tanganyika, Lindi District, Ruaha, *Busse* 1112 (B, holo. †, BM !, EA, iso.)
  C. burttii Exell in J.B. 77: 169 (1939).   Type: Zambia, 32 km. on Kasama–Isoka road, *B. D. Burtt* 5951 (BM, holo. !, EA, K !, iso.)
  C. mechowianum O. Hoffm. subsp. *taborense* (Engl.) Duvign. in B.S.B.B. 88: 80, t. 8/b (1956); Liben in F.C.B., Combr.: 76 (1968)
  [*C. mechowianum* sensu F.F.N.R.: 286 (1962), pro parte, *non* O. Hoffm. sensu stricto]

f. subsp. **suluense** (*Engl. & Diels*) *Okafor* in Bol. Soc. Brot., sér. 2, 41: 143 (1967).
Type: Swaziland, Horomine, *Galpin* 1264 (Z, holo., K !, PRE, iso.)

Young branches terete, lightly tomentose and lepidote. Leaf-lamina coriaceous, narrowly elliptic to oblong-elliptic, rarely broadly elliptic, up to 15(–22) cm. long and 10 cm. wide, apex generally acuminate, base cuneate or rounded, upper surface drying olive-brown, lower surface green- or yellow-brown, never silvery, glabrous (except for scales); scales (fig. 1/10i, p. 12) peltate, white, sometimes becoming yellowish, not contiguous; lateral nerves 8–9, generally prominent beneath and visible above; petiole 1–3(–4) cm. long, glabrous or lightly lepidote. Inflorescence simple. Lower receptacle glabrous (except for scales) and densely lepidote to densely pubescent (the hairs then concealing the scales); upper receptacle glabrous to lightly pubescent. Fruit (fig. 4/10b, p. 15) with conspicuous red scales, otherwise glabrous to slightly pubescent, wings with a blue- to yellow-brown " metallic " appearance.

KENYA. Machakos District: Yatta Plain, 29 Jan. 1938, *D. C. Edwards* 114 ! & Yatta
  Plateau, 15 Apr. 1958, *Trapnell* 2396 ! & S. of Makindu, 24 July 1953, *Trump* 75 !
TANGANYIKA. Bukoba District: Bukoba–Biharamulo road, Oct. 1967, *Procter* 732 !;
  Mpanda District: Mahali Mts., Silambula, 18 Sept. 1958, *Newbould & Jefford* 2081 !;
  Kilosa, 28 Oct. 1961, *Semsei* 3364 !
DISTR. **K**4; **T**1–8; Malawi, Mozambique, Zambia, Rhodesia, Lesotho, Angola, South
  West Africa, South Africa (Transvaal), Swaziland
HAB. *Brachystegia* woodland, wooded grassland and *Acacia, Commiphora* bushland;
  50–1700 m.

SYN. *? C. truncatum* Engl., P.O.A. C: 288 (1895), *non* Laws. (1871), *nom. illegit.*
        Type: Tanganyika, Tabora, *Stuhlmann* 572 (B, holo. †)
      C. fischeri Engl., P.O.A. C: 290 (1895); K.T.S.: 144 (1961). Types: Tanganyika,
        Mpanda District, Ugalla, Irunde, *Boehm* 19a (B, syn. †) & Dodoma District,
        Saranda, *Fischer* 246 (B, syn. †)
      C. brosigianum Engl. & Diels in N.B.G.B. 2: 192 (1898). Type: Tanganyika,
        Kilosa, *Brosig* (B, holo. †, BM, K, fragment !)
      ? C. oliveranum Engl., E.M. 3: 48, t. 14/C (1899); T.T.C.L.: 141 (1949). Type:
        as *C. truncatum* Engl.
      C. suluense Engl. & Diels in E.M. 3: 54, t. 14/G (1899); K.T.S.: 146 (1961)
      ? C. kerengense Diels in E.J. 39: 500 (1907); T.T.C.L.: 140 (1949). Type: Tangan-
        yika, Lushoto District, Kerenge–Kwashemshi, *Engler* 919A (B, holo. †)
      C. angustilanceolatum Engl., V.E. 3 (2): 702 (1921). Type: Mozambique, Lower
        Umswirizwi R., *Swynnerton* 45 (B, holo. †, BM, K !, iso.)
      C. makindense Engl., V.E. 3 (2): 703 (1921). Type: Kenya, Machakos District,
        Makindu, *Scheffler* 210 (B, holo. †, BM, K !, iso.)
      [*C. binderanum* sensu T.T.C.L.: 138 (1949); Liben in F.C.B., Combr.: 78 (1968),
        pro parte, *non* Kotschy]
      [*C. mechowianum* sensu F.F.N.R.: 286 (1962), pro parte, *non* O. Hoffm.]

NOTE. The identity of *C. oliveranum* and *C. kerengense* is not certain in the absence of
  authentic material, but from the descriptions they seem most likely to belong here.

Sect. 5. **Glabripetala** *Engl. & Diels* in E.M. 3: 43 (1899), emend. Exell &
Stace in Bol. Soc. Brot., sér. 2, 42: 22 (1968); Stace in J.L.S. 62: 158 (1969)

Flowers 4-merous. Leaves glutinous. Upper receptacle campanulate to infundibuliform at the base and cupuliform at the apex, clearly divided into 2 regions. Petals cuneate or spathulate to obovate, glabrous. Stamens 8, 1-seriate, inserted at the margin of the disk. Disk with a pilose free margin. Fruit 4-winged, glutinous. Cotyledons 2, arising below soil-level and borne above ground on a long stalk formed by the connate petioles. Scales 50–65 μ in diameter, whitish or yellowish, usually rather inconspicuous, circular in outline, not scalloped; cells delimited by 8(–13) radial walls alone; cell walls clear, cells transparent; more frequent on the upper epidermis of the leaf than on the lower.

Leaves glutinous, glabrescent in E. Africa; fruit
    subcircular to elliptic in outline, 2·5–3·5 cm.
    long, glutinous, rather inconspicuously lepi-
    dote, otherwise glabrous   .   .   .   .   11. *C. fragrans*

Leaves densely tomentose on the prominent
reticulation; fruit oblong-elliptic in outline,
3 cm. long, pubescent at first then glutinous . 12. *C. schweinfurthii*

11. **C. fragrans** *F. Hoffm.*, Beitr. Kenntn. Fl. Centr.-Ost-Afr.: 31 (1889);
P.O.A. C: 289 (1895); Engl. & Diels in E.M. 3: 51 (1899); T.T.C.L.: 140
(1949); Exell in Kirkia 7: 183 (1970). Types: Tanganyika, Mpanda District,
Ugalla, *Boehm* 32A (B, syn. †, Z, isosyn.!, K, fragment!) & Pa-Kabombue,
Boehm 16A (B, syn. †, Z, isosyn.!)

Small tree up to 10(–12) m. high; bark grey, reticulately fissured; branches
peeling to give dark reddish brown colour; branchlets glabrous, glutinous or
pubescent. Leaves opposite or 3(–4)-verticillate; lamina papyraceous to
coriaceous, narrowly to broadly ovate-elliptic to ovate, up to 15(–20) cm.
long and 9(–11) cm. wide, apex acute or obtuse, base usually cuneate, usually
glutinous, especially so when young, rarely tomentose in East Africa to
nearly glabrous (except for the scales), sparsely lepidote above, scales almost
contiguous beneath but often very difficult to see due to the glutinous exuda-
tion; lateral nerves 7–10(–13) pairs, usually prominent on both surfaces,
reticulation slightly raised beneath; petiole up to 1·5 cm. long, leaves sometimes subsessile, leaving a prominent circular scar when fallen, especially so
on the swollen nodes of the older wood. Inflorescences axillary, either single
simple spikes or clusters of such spikes borne on very much reduced axillary
shoots or single axillary spikes on elongated shoots (20 cm.) that are leafless
at flowering thus giving the appearance of a branched inflorescence, both
short and elongated shoots densely yellow-brown pubescent. Flowers (fig.
3/11, p. 14) greenish yellow to white becoming yellow, fragrant. Lower
receptacle usually tomentose; upper receptacle infundibuliform to broadly
campanulate at the base and cupuliform at the apex, 2–3 mm. long, 2–3 mm.
in diameter, pubescent to tomentose. Sepals broadly triangular. Petals
cuneate or spathulate to obovate, 2–3 mm. long, 1–1·5 mm. wide, glabrous.
Stamen-filaments 5–6 mm. long, inserted at the margin of the disk; anthers
0·8–1 mm. long. Disk with pilose free margin. Style 3–4 mm. long. Fruit
(fig. 4/11, p. 15) subcircular to elliptic in outline, 2·5–3·5 cm. long, 2·5–3 cm.
wide, yellow-brown to brown, glutinous, rather inconspicuously lepidote,
otherwise glabrous; apical peg up to 3 mm. long; wings up to 12 mm. broad;
stipe up to 5(–7) mm. long. Cotyledons 2, arising below soil-level and
borne above ground on a long stalk formed by the connate petioles. Scales
(fig. 2/11, p. 13) 50–65 μ in diameter.

UGANDA. W. Nile District: Rhino Camp, 27 Mar. 1936, *Michelmore* 1395!; Karamoja
  District: Namalu, 2 Dec. 1958, *Langdale-Brown* 84!; Teso District: Serere, *Eggeling*
  759 in *Brasnett* 1153!
KENYA. Central Kavirondo/Kisumu District: escarpment above Kano Plain, near
  Songhor, 28 Jan. 1964, *Brunt* 1428!; Masai District: Kilaguni, 1 July 1968, *V. C.
  Gilbert* 2761!
TANGANYIKA. Shinyanga District: Mwantine Hills, Oct. 1935, *B. D. Burtt* 5292!;
  Tanga/Pangani District: Songa [Mabatini], 14 Sept. 1960, *Paulo* 780!; Tabora
  District: Itulu Hill, Sept. 1951, *Groome* 22!
DISTR. U1, 3, 4; K5, 6; T1, 3–8; W. Africa, extending to Zaire and Sudan, also Zambia,
  Rhodesia, Malawi, Mozambique and Botswana
HAB. Common in deciduous woodland and wooded grassland, often associated with
  seasonally waterlogged clay soils, but sometimes on shallow stony soils; 50–1700 m.

SYN. [*C. reticulatum* sensu Laws. in F.T.A. 2: 432 (1871), pro parte quoad specim.
  Grant 734.5; Oliv. in Trans. Linn. Soc. 29:71 (1873); I.T.U., ed. 2: 85 (1952),
  non Presl (1827), *nec* Fresen. (1837)]
  ? *C. kilossanum* Engl. & Diels in N.B.G.B. 2: 193 (1898); T.T.C.L.: 140 (1949).
    Type: Tanganyika, Kilosa, *Brosig* (B, holo. †)
  ? *C. albidiflorum* Engl. & Diels in E.M. 3: 46, t. 14/A (1899); T.T.C.L.: 140 (1949).
    Type: Tanganyika, foothills of Uluguru Mts., *Stuhlmann* 8991 (B, holo. †,
    BM, fragment!)

C. ghasalense Engl. & Diels in E.M. 3 : 47, t. 15/B (1899); F.P.S. 1 : 206 (1950); I.T.U., ed. 2 : 86 (1952); K.T.S.: 145 (1961); Liben in F.C.B., Combr.: 75 (1968). Types: Sudan, Bahr el Ghazal, Sabbi [Ssabi], 1869, *Schweinfurth* 2730 & Bahr el Ghazal, R. Tuju [Tudje], 1869, *Schweinfurth* 2745 (both B, syn. †, BM, fragments !, K, isosyn. !)

C. multispicatum Engl. & Diels in E.M. 3 : 47, t. 15/A (1899); F.P.S. 1 : 204 (1950); Liben in F.C.B., Combr.: 75 (1968). Type: Sudan, Bahr el Ghazal, R. Tuju–Sabbi [Tudje–Ssabi], 1869, *Schweinfurth* 2662 (B, holo. †, BM, fragment !, K, iso. !)

C. subvernicosum Engl. & Diels in E.M. 3 : 48 (1899); T.T.C.L.: 141 (1949). Type: Tanganyika, Buha/Kigoma Districts, Malagarasi, *von Trotha* 34 (B, holo. †)

C. undulatum Engl. & Diels in E.M. 3 : 48, t. 15/C (1899); F.P.S. 1 : 204 (1950). Types: Sudan, Bahr el Ghazal, *Schweinfurth* 1306, 1511, 2802A & 2815 (B, syn. †, BM, fragment of 2815 !, K, isosyn. of 2815 !)

C. ternifolium Engl. & Diels in E.M. 3 : 49, t. 14/D (1899); T.T.C.L.: 141 (1949). Type: Tanganyika, Morogoro District, Mgeta, *Stuhlmann* 9272 (B, holo. †, BM, K !, fragments)

C. tetraphyllum Diels in E.J. 39: 499 (1907); T.T.C.L.: 141 (1949). Type: Rhodesia/Zambia, Victoria Falls, *Engler* 2916 (B, holo. †, BM, fragment !)

C. sp. near C. ghasalense sensu Burtt Davy, Check-lists Brit. Emp. 1, Uganda: 36 (1935)

C. sp. sensu Burtt Davy, Check-lists Brit. Emp. 1, Uganda: 37 (1935), pro specim. *Chandler* 178 ! & 398 !

NOTE.   The identity of *C. albidiflorum* and *C. kilossanum* is not certain in the absence of authentic material, but from the descriptions they seem most likely to belong here.

12. **C. schweinfurthii** *Engl. & Diels* in E.M. 3 : 50, t. 14/E (1899); Burtt Davy, Check-lists Brit. Emp. 1, Uganda: 36 (1935); F.P.S. 1 : 203 (1950); I.T.U., ed. 2 : 88 (1952); Liben in F.C.B., Combr.: 74 (1968). Types: Sudan, Equatoria, Gumango Hill, 1870, *Schweinfurth* 2886 & Bahr el Ghazal, Ngoli, 1870, *Schweinfurth* 2903 (both B, syn. †, K, isosyn. !)

Small tree or bush up to 3 m. high; bark grey, longitudinally fissured; young branches yellow-brown pubescent. Leaves opposite or 3-verticillate; lamina coriaceous, oblong-ovate, ovate or obovate, up to 35 cm. long, 9–13 cm. wide on mature trees (more information is needed to verify these observations), apex obtuse or subacute, base shortly attenuate, margin usually slightly sinuate; upper surface drying olive-brown, subglutinous, glabrescent except for the yellow-brown indumentum on the prominent midrib and lateral nerves; lower surface yellow-brown with light brown venation, venation with white to buff tomentum obscuring the sparse scales in the areoles; midrib, lateral nerves and reticulation strongly prominent beneath, laterals 8–20 pairs; petiole stout, up to 3·5 cm. long, leaving large sunken scars at the swollen nodes of the older branches. Inflorescences axillary clusters of brown tomentose racemes 9–11 cm. long, usually 6 in number, sometimes borne along a short (2–3 cm. long) tomentose axillary shoot. Lower receptacle 2–2·5 mm. long, velutinous-tomentose; upper receptacle cupular, 3–4 mm. long, 3–4 mm. in diameter, pubescence decreasing in density and becoming greyer towards the shallowly triangular sepals. Petals broadly obovate, 2–2·5 mm. long, 1·5–2 mm. wide, glabrous. Stamen-filaments slender, 5–6 mm. long; anthers 1 mm. long. Disk 2 mm. in diameter, with long-pilose margin free for 0·5 mm. Style up to 6 mm. long. Fruit (fig. 4/12, p. 15) oblong-elliptic in outline, 3 cm. long, 2·3 cm. wide, pubescent at first, then glutinous, eventually wings a silky yellow-brown with darker shortly pubescent and glutinous centre; stipe stout, 2–3 mm. long, pubescent. Scales (fig. 2/12, p. 13)   50–65 μ in diameter.

UGANDA.   W. Nile District: Okollo, Feb. 1934, *Eggeling* 1543 ! & 1974 ! & Meturu, near Arua, *Eggeling* 1778 !
DISTR.   U1 ; Zaire and Sudan
HAB.   Probably wooded grassland; 750–1200 m.

NOTE. This species is insufficiently known. Very little flowering material is available and little is known of its growth habit. It is reported to be common near Okollo in West Nile District. Its unusual *Terminalia*-like leaves make it easy to recognize and should facilitate further collections and observations. There would appear to be some affinities between *C. schweinfurthii* and *C. gallabatense* Schweinf. from the Sudan and Ethiopia, but *C. gallabatense* too has been poorly collected. Its affinities with *C. fragrans* will require further study when more material becomes available.

    Greenway & Eggeling 7243, from W. Nile District, Koka to Metu, appears to be an aberrant form, with smaller and narrow leaves, approaching *C. gallabatense* in shape but not in indumentum.

Sect. 6. **Spathulipetala** *Engl. & Diels* in E.M. 3 : 58 (1899); Stace in J.L.S. 62 : 158 (1969)

    Flowers 4-merous. Upper receptacle shortly infundibuliform. Petals obovate-spathulate to spathulate, glabrous. Stamens 8, 1-seriate, inserted at the margin of the disk. Disk with pilose margins free for 1·5–2 mm. Style with swollen apex (while functional), usually exserted before the stamens. Fruit 4-winged, large for the genus, up to 6·5(–10) cm. long with stipe up to 2·5 cm. long. Cotyledons 2, arising below soil-level and borne above soil on a long stalk formed by the connate petioles, sometimes united to form a single peltate organ. Scales (fig. 2/13, p. 13) ± 40–75 μ in diameter, whitish or yellowish, circular but slightly convex, cells delimited by 7–9 radial walls and also by a few tangential and extra radial walls, marginal cells 7–12; cell walls clear and cells transparent.

Only species   .   .   .   .   .   .   13. *C. zeyheri*

    13. **C. zeyheri** *Sond.* in Linnaea 23 : 46 (1850); Engl. & Diels in E.M. 3 : 59 (1899); T.T.C.L.: 140 (1949); K.T.S.: 147 (1961); F.F.N.R.: 285 (1962); Liben in F.C.B., Combr.: 72 (1968); Exell in Kirkia 7 : 185 (1970). Type: South Africa, Transvaal, Magaliesberg, *Zeyher* 552 (TCD, holo., B †, K !, MEL, Z, iso.)

    Small to medium-sized tree up to 10(–13) m. high or rarely a shrub; bark brown or grey-brown, tessellated; branchlets light brown, pubescent. Leaves opposite or 3-verticillate; lamina chartaceous, broadly to narrowly elliptic to obovate-elliptic or oblong-elliptic, up to 14(–16) cm. long and 8(–10) cm. wide, apex usually rounded to obtuse, sometimes acute, base usually rounded, sometimes slightly cordate, usually tomentose (when young) or pubescent to almost glabrous (except for the scales), lepidote, but scales small and rather inconspicuous; lateral nerves 5–12 pairs, somewhat prominent beneath; petiole up to 1 cm. long. Inflorescences usually unbranched axillary spikes, up to 8 cm. long including the rather tomentose peduncle. Flowers (fig. 3/13, p. 14) greenish yellow; odour, if any, not recorded. Lower receptacle tomentose; upper receptacle shortly infundibuliform, ± 2·5–3 mm. long, 3 mm. in diameter, pubescent and lepidote. Sepals triangular. Petals obovate-spathulate to spathulate, 1·5–3 mm. long, 0·8–1·2 mm. wide, glabrous. Stamen-filaments 5–8 mm. long, often reflexed ± 1 mm. below the anthers; anthers buff to orange, 1·5 mm. long. Disk with pilose margin free for 1·5–2 mm. Style precocious, 2·5 mm. long, swollen at the apex. Fruit (fig. 4/13, p. 15) subcircular or elliptic to transversely elliptic (not in E. Africa), usually 6·5 cm. long by 5·5–6 cm. wide (rarely only 3–5 cm. long, 3–4 cm. wide, exceptionally up to 10 cm. long, 8 cm. wide), straw coloured to light brown, usually glabrescent, sometimes conspicuously lepidote on the body; apical peg very short or absent; wings up to 4 cm. wide; stipe 1–3 cm. long, usually relatively slender. Cotyledons and scales (fig. 2/13) as for the section.

KENYA. Machakos District: Sultan Hamud, 20 Sept. 1953, *Drummond & Hemsley* 4427! & Machakos–Kitui road, Sept. 1960, *Dale* K 1089! & Makueni, 17 Oct. 1947, *Bogdan* 1358!

TANGANYIKA. Shinyanga District: 22 May 1956, *Gane* 66!; Dodoma District: Manyoni Kopje, 27 July 1933, *B. D. Burtt* 4969!; Kilwa District: Kilwa Kivinje, 4 Dec. 1955, *Milne-Redhead & Taylor* 7552!

DISTR. K4; T1–8; Zaire, Zambia, Malawi, Mozambique, Rhodesia, Angola, Botswana, South West Africa and South Africa (Natal, Transvaal)

HAB. *Brachystegia* woodland, wooded grassland and *Acacia*, *Commiphora* bushland, usually on sandy soils, also on termite mounds; 10–1600 m.

SYN. *C. glandulosum* F. Hoffm., Beitr. Kenntn. Fl. Centr.-Ost-Afr.: 33 (1889); P.O.A. C: 289 (1895); T.T.C.L.: 140 (1949). Type: Tanganyika, Tabora District, Igonda [Gonda], *Boehm* 155a (B, holo. †, BM, fragment!)
     C. oblongum F. Hoffm., Beitr. Kenntn. Fl. Centr.-Ost-Afr.: 34 (1889); P.O.A. C: 289 (1895). Type: Tanganyika, Tabora District, Kakoma, *Boehm* 87a (B, holo. †, K, fragment!)
     C. platycarpum Engl. & Diels in E.J. 39: 503 (1907). Type: Tanganyika, Kilwa District, Donde–Barikiwa [Barigiwa], *Busse* 586 (B, holo. †, EA, iso.!)
     C. rupicolum Diels in E.J. 39: 503 (1907), *non* Ridl. (1890), *nom. illegit.* Type: Tanganyika, Dodoma District, Kilimatinde, *Busse* 245 (B, holo. †)
     C. megalocarpum Brenan, T.T.C.L.: 140 (1949). Type: as *C. rupicolum* Diels

Sect 7. **Ciliatipetala** *Engl. & Diels* in E.M. 3: 32 (1899), excl. *C. fulvoto-mentosum* Engl. & Diels; Stace in J.L.S. 62: 149 (1969)

Flowers 4-merous. Upper receptacle cupuliform to broadly campanulate. Petals small (sometimes minute or even absent), obovate or obcuneate to subcircular, sometimes emarginate, ciliate or pilose at the apex (very rarely glabrous as in *C. psidioides* subsp. *glabrum*—not in E. Africa). Stamens 8, 1-seriate in insertion or nearly so. Disk with a short free pilose margin. Fruit 4-winged. Cotyledons 2, with long petioles arising at or below soil-level or (*C. albopunctatum*—not in E. Africa) with short petioles borne above soil-level. Scales somewhat variable, 40–120(–130) μ in diameter, ± circular, slightly or markedly scalloped at the margin, delimited by (7–)8– ± 12 primary radial walls and often with additional tangential walls. The scale characters in the section are more heterogeneous than for the other sections of this genus, and can be separated out into a number of scale types. The scales are the only character that can satisfactorily distinguish between some of the glabrous forms of *C. molle* and *C. apiculatum*, an extremely unsatisfactory state of affairs.

Bark of young branches coming off in untidy fibrous strips or threads:
     Reticulation of lower surface of leaf usually prominent, sometimes subprominent; leaves up to 14(–21) cm. long, typically sericeous tomentose, but sometimes glabrous; fruit rarely exceeding 2 cm. in length or width; scales (fig. 2/14, p. 13) ± (75–)90–120(–130) μ in diameter, cells opaque with walls appearing thick; a small tree or shrub   .   14. *C. molle*
     Reticulation of lower surface of leaf usually not prominent, sometimes subprominent in intermediates; leaves up to 14 cm. long, glutinous when young; fruit up to 3 cm. long and 2·7 cm. wide, usually glutinous; scales (45–)50–75(–100) μ in diameter, transparent and thin-walled:
       Small tree or shrub; leaves usually glabrous on the reticulation beneath, apiculate   .   15. *C. apiculatum*

Climber or scrambler; leaves remaining sparsely
    hairy at maturity and pilose around the
    margins, apex acute to acuminate .   .   16. *C. acutifolium*
Bark of branchlets peeling off in large ± cylindrical
    or hemi-cylindrical pieces leaving an exposed
    cinnamon-red surface; reticulation of lower
    leaf-surface prominent; scales 10–60(–70) μ in
    diameter, thin-walled and obscured by glutinous
    secretions .   .   .   .   17. *C. psidioides*

14. **C. molle** *G. Don* in Trans. Linn. Soc. 15: 431 (1827) [R.Br. in Salt,
Voy. Abyss. App.: lxiv (1814), *nom. nud.*]; T.T.C.L.: 137 (1949); F.P.S.:
201 (1950); I.T.U., ed. 2: 88 (1952); K.T.S.: 145 (1961); Liben in F.C.B.,
Combr.: 67 (1968); Exell in Kirkia 7: 189 (1970). Type: Ethiopia, without
precise locality, *Salt* (BM, holo.!)

Generally a small tree 5–7 m. high, sometimes 10–17 m. high, sometimes
shrubby; bark dark grey to black, rough, reticulately fissured (resembling
crocodile skin); branchlets with bark peeling in grey fibrous strips. Leaves
opposite (sometimes 3-verticillate in coppice regrowth); lamina generally
coriaceous, sometimes papyraceous (variant "A"), narrowly elliptic or narrowly ovate-elliptic or ovate-triangular or broadly ovate-elliptic or obovate or
obovate-elliptic, up to 21 cm. long and 12·5 cm. wide, apex acute or obtuse,
base cuneate or rounded or cordate, typically pubescent above and densely
grey tomentose beneath, sometimes indumentum a dark velvety brown on
the old leaves and pinkish on the young leaves, some forms almost glabrous
above and pubescent beneath or glabrous on both surfaces (except for the
scales); scales silvery or reddish, contiguous or almost so, but often hidden
by the indumentum; lateral nerves 6–12 pairs, the parallel reticulation usually
very prominent beneath (but less so in apparently transitional forms);
petiole usually 2–3 mm. long. Inflorescences of axillary spikes up to 7(–11)
cm. long, occasionally forming panicles by suppression of the upper leaves;
peduncles 1–2 cm. long, pubescent. Flowers yellow or greenish yellow,
fragrant. Lower receptacle tomentose; upper receptacle campanulate,
1·5–3 mm. long, 2–3 mm. in diameter, tomentose, lepidote. Sepals broadly
deltate. Petals sometimes minute or absent, irregularly obovate-deltate to
reniform, 0·5–1 mm. long and wide, apex ciliate. Stamen-filaments 5–6 mm.
long, inserted at the margin of the disk; anthers 1 mm. long. Disk with a
very short pilose free margin less than 0·5 mm. wide. Style 3–5 mm. long.
Fruit (fig. 5/14, p. 16) subcircular to elliptic in outline, 1·3–2·3(–2·5) cm. long,
1·5–2·3(–2·5) cm. wide, pale straw coloured to yellow-brown, lepidote and
tomentose (especially on the body) to nearly glabrous (except for the scales);
wings papyraceous (especially variant " A "); apical peg up to 1 mm. long;
stipe 2–5 mm. long. Cotyledons 2, arising at or below soil-level. Scales
(fig. 2/14, p. 13) discoid with marginal cells usually concave, ± (75–)90–120
(–130) μ in diameter, usually with 8 radial walls and typically 8 (rarely 0)
tangential walls and sometimes with many to several extra radial walls;
marginal cells 8–11(–16); cell walls clear and usually very thick; cells usually
opaque.

UGANDA. Kigezi District: Bugangari, Feb. 1950, *Purseglove* 3264!; Elgon, 5 Feb. 1964,
    E.S. Brown 755!; Masaka District: Kabula, Oct. 1932, *Eggeling* 689!
KENYA. Nairobi District: Bahati, 10 Mar. 1933, *C. G. Rogers* 555! & Nairobi, 8 Oct.
    1934, *Napier* 3550 in *C.M.* 6636!; Masai District: Ngong Hills, 3 Sept. 1965, *Kokwaro,
    Kanuri & Mathenge* 319!
TANGANYIKA. Singida District: Mkalama, Mshaushi, 22 Oct. 1925, *B. D. Burtt* 5338!;
    Dodoma District: Itigi, 9 Apr. 1964, *Greenway & Polhill* 11465!; Morogoro District:
    without precise locality, Sept. 1952, *Semsei* 935!

DISTR.   U1–4; K1–7; T1–8; throughout the wooded grassland areas of tropical and
southern Africa; also in the Yemen (the only African species found outside Africa)
HAB.   Widespread throughout the wooded grassland and bushland areas of East Africa,
often forming pure stands on hillsides; 30–2300 m.

SYN.   *C. trichanthum* Fresen. in Mus. Senckenb. 2: 155 (1837); Engl. & Diels in E.M.
3: 34, t. 9/B (1899); Burtt Davy, Check-lists Brit. Emp. 1, Uganda: 36 (1935).
Type: Ethiopia, Gondar–Adua, *Rueppell* (FR, holo.)
    C. ferrugineum A. Rich., Tent. Fl. Abyss. 1: 267 (1847); Burtt Davy, Check-lists
Brit. Emp. 1, Uganda: 35 (1935). Type: Ethiopia, Selassaquilla, *Schimper* 767
(P, holo., K, iso. !)
    C. petitianum A. Rich., Tent. Fl. Abyss. 1: 268 (1847); Burtt Davy, Check-lists
Brit. Emp. 1, Uganda: 36 (1935). Type: Ethiopia, Shoa [Choa], *Petit* (P, holo.)
    C. gueinzii Sond. in Linnaea 23: 43 (1850); Engl. & Diels in E.M. 3: 38, t. 12/A
(1899); Burtt Davy, Check-lists Brit. Emp. 1, Uganda: 35 (1935); I.T.U., ed.
2: 88 (1952). Type: South Africa, Natal, Durban [Port Natal], *Gueinzius* 567
(S, holo., K, iso. !)
    C. gondense F. Hoffm., Beitr. Kenntn. Fl. Centr.-Ost-Afr.: 32 (1889); P.O.A. C:
291 (1895); Engl. & Diels in E.M. 3: 55 (1899); T.T.C.L.: 138 (1949). Type:
Tanganyika, Tabora District, Igonda [Gonda], *Boehm* 66A (B, holo. †, BM,
fragment !)
    C. splendens Engl., P.O.A. C: 289 (1895); Engl. & Diels in E.M. 3: 37, t. 11/D
(1899); Burtt Davy, Check-lists Brit. Emp. 1, Uganda: 36 (1935); T.S.K., ed.
2: 32 (1936). Types: Tanganyika, Dodoma District, Saranda, *Fischer* 248, 249
& 251 (B, syn. †) & Shinyanga District, Salawe, *Stuhlmann* 699 (B, syn. †) &
Malawi, without precise locality, *Buchanan* 859 (B, syn. †, K, isosyn. !)
    C. deserti Engl., P.O.A. C: 289 (1895); Engl. & Diels in E.M. 3: 35, t. 11/C (1899);
Diels in E.J. 39: 493 (1907); T.T.C.L.: 136 (1949); K.T.S.: 144 (1961). Types:
Kenya, Kwale District, Maruvessa, *Hildebrandt* 2362 (B, syn. †, BM, fragment !,
K, isosyn. !) & Tanganyika, Moshi District, Marangu, *Volkens* 2126 (B, syn. †,
BM, isosyn. !) & Bukoba District, Ngono R. [Kjanjaviassi], *Stuhlmann* 3229
(B, syn. †)
    C. nyikae Engl. [in Abh. Preuss. Akad. Wiss. 1894: 36 (1894), nom. nud.],
P.O.A. C: 289 (1895). Types: Tanganyika, Lushoto/Tanga District, Hosiga,
Holst 2539 (B, syn. †, K, isosyn. !) & Moshi District, Marangu, *Volkens* 1710
(B, syn. †, BM, isosyn. !) & Mashame [Madschame], *Volkens* 1662 (B, syn. †,
BM, K, isosyn. !)
    C. nyikae Engl. var. *boehmii* Engl., P.O.A. C: 290 (1895). Types: Tanganyika,
Tabora/Mpanda District, Ugalla, *Boehm* 125a (B, syn. †) & Lushoto District,
Mashewa, *Holst* 8855 (B, syn. †, K, isosyn. !)
    C. microlepidotum Engl., P.O.A. C: 290 (1895); Engl. & Diels in E.M. 3: 35, t.
10/C (1899); T.T.C.L.: 137 (1949); K.T.S.: 145 (1961). Types: Tanganyika,
Mwanza District, Busisi [Bussisi], *Stuhlmann* 766 (B, syn. †) & *Stuhlmann* 774
(B, syn. †, K, fragment !)
    C. schelei Engl., P.O.A. C: 291 (1895). Type: Tanganyika, Tanga District,
Amboni, *Holst* 2916 (B, holo. †, K, iso. !)
    C. tenuispicatum Engl., P.O.A. C: 291 (1895); Engl. & Diels in E.M. 3: 41,
t. 10/D (1899); T.T.C.L.: 136 (1949). Types: Tanganyika, Lushoto District,
Gombero [Gombelo], *Holst* 2165 (B, syn. †, K !, Z, isosyn.) & Morogoro District,
Usagara, Vilanzi R., *Stuhlmann* 88 (B, syn. †)
    C. ulugurense Engl. & Diels in E.M. 3: 35, t. 10/A (1899); T.T.C.L.: 137 (1949).
Type: Tanganyika, Morogoro District, Tununguo, *Stuhlmann* 8948 (B, holo. †,
BM, fragment !)
    C. hobol Engl. & Diels in E.M. 3: 36, t. 11/A (1899). Type: Somali Republic (N.),
Sogair–Berbera, *Bricchetti* 25 (B, holo. †, K, fragment !, RO, iso.)
    C. splendens Engl. var. *nyikae* (Engl.) Engl. (incl. *C. nyikae* Engl. var. *boehmii*
Engl.), E.M. 3: 37, t. 11/E (1899)
    C. welwitschii Engl. & Diels in E.M. 3: 40 (1899); Burtt Davy, Check-lists Brit.
Emp. 1, Uganda: 36 (1935). Type: Angola, Cuanza Norte, Golungo Alto,
Welwitsch 4318 (LISU, lecto., B †, BM !, COI, K !, P, isolecto., selected by Exell
in C.F.A. 4: 63 (1970))
    C. holtzii Diels in E.J. 39: 494 (1907); T.T.C.L.: 137 (1949). Types: Tanganyika,
Tabora, *Holtz* 1478 & 1518 (B, syn. †)
    C. ankolense Bagshawe & Bak. f. in J.B. 46: 4 (1908); Burtt Davy, Check-lists
Brit. Emp. 1, Uganda: 35 (1935). Type: Uganda, Ankole District, Mulema,
Bagshawe 212 (BM, holo. !)
    C. sp. sensu Burtt Davy, Check-lists Brit. Emp. 1, Uganda: 37 (1935), pro
specim. *Liebenberg* 911 ! & *F.D.* 1006 !
    C. gueinzii Sond. subsp. *splendens* (Engl.) Brenan, T.T.C.L.: 137 (1949)
    C. sp. aff. C. microlepidotum Engl. sensu I.T.U., ed. 2: 89 (1952)

NOTE. Some further synonymy will be found in Exell & Garcia in Contr. Conhec. Fl. Moçamb. 2: 106 (1954) and Exell in Kirkia 7: 189 (1970).

This is an extremely variable species, varying chiefly in leaf-shape, reticulation and indumentum and in the size of the fruits. These characters appear in various combinations throughout the distribution range and further confusion is caused through the presence of intermediate forms. It is possible that lengthy research will enable a number of subspecies to be established, but until this can be done satisfactorily throughout its entire range, it is necessary to treat the whole as an aggregate species. This is a necessary but unsatisfactory method, especially for the field worker who may be able to recognize distinct taxa within his own geographical area. In E. Africa it is possible to recognize three major divisions within the aggregate species.

The first forms by far the main bulk of the herbarium specimens and consists of plants having relatively large coriaceous leaves and variable fruits. The second, variant " A " from **T1**, is characterized by leaves up to 11 cm. long and 5 cm. wide, ovate-elliptic, apex acute, base cordate, yellow-green, papyraceous, lateral nerves 7–10 pairs, reticulation not prominent, pilose only on the venation; fruit (fig. 5/14a, p. 16) small, 2 cm. long, 1·8 cm. wide, wings papyraceous, silvery or straw coloured. Specimens: **T1**, Shinyanga District, granite hills, *B. D. Burtt* 2502, 2919, 5054, 5108, 5109, 5297, 5559, 6461 & 6532. *Bally* 12337 from **K3**, Kerio valley, may also belong here. The variant is related to the *C. trichanthum-C. hobol* group within the aggregate species, which occurs in Ethiopia and has somewhat similar fruits and even more glabrescent leaves.

In the dry country of north-eastern Africa, south through **K1, 4, 7, T3** to **T4/5**, there is a further variant, variant " B ", characterized by a straggling habit, relatively small elliptic leaves (up to 5·5 cm. long and 2·5 cm. wide), which may be densely hairy but often only lightly pubescent, the venation scarcely prominent, rather like variant " A ", but the fruits more like the main part of *C. molle*. *C. deserti* (see synonymy above) belongs here. Specimens: **K1**, Dandu, *Gillett* 12561 & 12562; **K4**, Kitui District, Ndui Rock Dam, *Kimani* 109; **T4/5**, Wembere Escarpment, *Richards* 13436 & 13451. A few plants of variant " B " approach *C. apiculatum*.

15. **C. apiculatum** *Sond.* in Linnaea 23: 45 (1850); Engl. & Diels in E.M. 3: 42, t. 12/C (1899); T.T.C.L.: 136 (1949); K.T.S.: 141 (1961); Exell in Kirkia 7: 192 (1970). Type: South Africa, Transvaal, Magaliesberg, *Zeyher* 553 (S, holo., K, iso.!)

Small tree up to 10 m. high, rarely a shrub; bark grey to greyish black, smooth or reticulated; leaf-buds black or dark brown. Leaves opposite; lamina subcoriaceous to papyraceous, glutinous when young, broadly to narrowly obovate-elliptic or oblong-elliptic or ovate to subcircular, 3–14 cm. long, 1·5–7·5 cm. wide, apex usually apiculate or mucronate and usually twisted, base usually rounded to subcordate, mature leaves usually glabrous (except for scales) in E. Africa, sometimes pilose along the midrib and principal nerves beneath, then lamina less than 7 cm. long; scales yellowish red, not contiguous and sometimes inconspicuous due to the glutinous secretions; lateral nerves 5–10 pairs, reticulation rather inconspicuous to prominulous beneath; petiole up to 10 mm. long. Inflorescences of axillary spikes up to 3–7 cm. long. Flowers yellow, fragrance if any, not recorded. Lower receptacle lepidote and otherwise hairy or glabrous; upper receptacle campanulate, 3–4 mm. long, 2·5–3 mm. in diameter, lepidote and otherwise glabrous or pubescent. Sepals broadly deltate. Petals obtriangular, 1·2 mm. long and wide, apex ciliate. Stamen-filaments 5 mm. long, inserted at the margin of the disk; anthers 1·2–1·5 mm. long. Disk with a pilose scarcely free margin. Style 4 mm. long. Fruit (fig. 5/15, p. 16) subcircular to elliptic or oblong-elliptic in outline, 2–3 cm. long, 1·5–2·5 cm. wide, glutinous when young, shiny when mature, lepidote and otherwise glabrous or pubescent; apical peg up to 1 mm. long; wings up to 7 mm. wide; stipe 4–8 mm. long. Cotyledons 2, arising below soil-level; petiole 4–5 cm. long. Scales (fig. 2/15, p. 13) 45–75(–90) μ in diameter, scalloped at each marginal cell; cells delimited by (7–)8(–10) radial walls and sometimes also up to 5 tangential walls and rarely with extra radial walls, marginal cells 7–10; cell walls clear and thin, cells not opaque.

subsp. **apiculatum**; Exell in Mitt. Bot. Staats. Münch. 4 : 3 (1961); Stace in Mitt. Bot. Staats. Münch. 4 : 13, t. 2/16a (1961); Exell in Kirkia 7 : 194 (1970)

Leaf-lamina up to 11–14 cm. long, 5–7·5 cm. wide, glabrous (except for scales) when mature except for tufts of hairs in the axils of the nerves beneath and occasional pilosity along the midrib and some of the principal nerves.

KENYA. Northern Frontier Province : Ol Doinyo Lengio, 19 Jan. 1958, *Newbould* 3539 !; Machakos District : Makueni, 17 Oct. 1947, *Bogdan* 1357 ! & Apr. 1958, *Trapnell* 2394 !
TANGANYIKA. Mbulu District : SW. of Lake Eyasi, 8 Jan. 1967, *Herlocker* 574 !; Dodoma District : Kilimatinde, 5 Sept. 1931, *B. D. Burtt* 3391 !; Morogoro District : Kisaki, 6 Dec. 1933, *B. D. Burtt* 5020 !
DISTR. K1, 4, 7; T1–8; Botswana, Zambia, Rhodesia, Malawi, Mozambique, Angola, South and South West Africa
HAB. *Brachystegia* woodland, wooded grassland and *Acacia*, *Commiphora* bushland, common on rocky hill slopes; 70–1800 m.

SYN. *C. buchananii* Engl. & Diels in E.M. 3 : 40 (1899). Type : Malawi, without precise locality, *Buchanan* 1263 (B, holo. †, K, iso. !)
  C. ukamense Engl. & Diels in E.M. 3 : 57 (1899); Diels in E.J. 39 : 502 (1907). Type : Tanganyika, Morogoro District, Ukami, *Stuhlmann* 8162 (B, holo. †, BM, fragment of holo. !)
  C. apiculatum Sond. subsp. *boreale* Exell in J.B. 67 : 46 (1929); T.T.C.L. : 136 (1949). Type : Tanganyika, Kondoa District, Kandaga, *B. D. Burtt* 1224 (BM, holo. !, K, iso. !)

NOTE. Subsp. *leutweinii* (Schinz) Exell, with generally smaller leaf-blades densely to sparsely pubescent on both surfaces and often rufous tomentose on the nerves, may occur in T8 on the Mozambique border, but no specimens have been seen. Some specimens from the drier parts of coastal Kenya (K7) approach subsp. *leutweinii*, but in view of the gap in distribution, the subspecies otherwise occurring only in Mozambique, Malawi and Zambia south to Rhodesia and South West Africa, I prefer to regard these specimens as ecological variants of subsp. *apiculatum*.

16. **C. acutifolium** *Exell* in J.B. 71, Suppl. : 232 (1933); F.F.N.R. : 286 (1962); Exell in Kirkia 7 : 195 (1970). Type : Angola, Bié, *Gossweiler* (BM, holo.!)

Slender-stemmed climber or scandent shrub. Leaves opposite; lamina papyraceous, elliptic or broadly elliptic, up to 7 cm. long, 3·5 cm. wide, apex acuminate and slightly mucronate, base rounded, nerves on both surfaces rather sparsely pubescent to pilose, sparsely pilose on the areoles beneath, densely ciliate on the margin, rather inconspicuously lepidote; lateral nerves 5–6 pairs; petiole up to 1·5 cm. long. Inflorescences of unbranched axillary spikes up to 7 cm. long. Flowers yellow. Lower receptacle glabrous or nearly so; upper receptacle broadly campanulate, ± 1·5 mm. in diameter, 2 mm. long, sparsely and rather inconspicuously lepidote, otherwise glabrous or nearly so. Sepals scarcely developed. Petals subcircular, shortly clawed, 1 mm. long and wide, apex ciliate. Stamen-filaments 5 mm. long; anthers 0·5–0·6 mm. long. Disk with short free pilose margin. Fruit (fig. 5/16, p. 16) subcircular in outline, up to 2·5 cm. long, 2·7 cm. wide, greenish yellow; wings sometimes red, ± 10 mm. broad, glutinous, ± glabrous except for scales; apical peg short or absent; stipe up to 7 mm. long. Scales (fig. 2/16, p. 13) 50–90(–100) μ in diameter, usually 8-celled with additional tangential and/or radial walls.

TANGANYIKA. Ufipa District : Kalambo Falls, 26 Mar. 1960, *Richards* 12802 !; Dodoma District : Luwila [Ruwiri] ravine, 12 Dec. 1935, *B. D. Burtt* 5387 !
DISTR. T4, 5; Zambia and Angola
HAB. Riverine forest and thicket, sometimes on termite mounds; 1200–1500 m.

NOTE. Very near *C. apiculatum* but distinguished by its scandent habit, characteristic fringe of hairs around the edges of the young leaves, longer petiole and smaller flowers.

17. **C. psidioides** *Welw.* in Ann. Cons. Ultram. parte não off., sér. 1, 1856 : 249 (1856); Engl. & Diels in E.M. 3 : 51 (1899); Exell in Mitt. Bot. Staats. Münch. 4 : 3 (1961); Liben in F.C.B., Combr.: 66 (1968); Exell in Kirkia 7 : 196 (1970); Wickens in K.B. 26 : 37 (1971). Type: Angola, Quicuxe, *Welwitsch* 4378 (LISU, lecto., BM !, COI, K !, isolecto.)

Tree up to 17 m. high or large bush; bark grey, deeply rectangularly fissured or fairly smooth; branchlets usually tomentose when young, with the bark peeling off in large dark grey or purple-black strips or cylindrical or hemicylindrical pieces, leaving a newly exposed cinnamon-red surface. Leaves opposite; lamina coriaceous, elliptic or oblong-elliptic or ovate-elliptic or ovate-oblong, 5–15(–24) cm. long, 3–10(–16) cm. wide, apex usually rounded or retuse and often mucronate or apiculate, base rounded or subcordate, densely ferruginous tomentose when young (except in subsp. *psilophyllum* where the young leaves are markedly glutinous), eventually white pubescent to pilose or tomentose on the reticulation beneath and glabrous on the areoles (subsp. *psidioides*) or ferruginous pubescent along the sides of the midrib and laterals (subsp. *psilophyllum*) or glabrous except for the scales (subsp. *glabrum*); scales reddish, widely scattered, inconspicuous; lateral nerves 8–16 pairs, reticulation rather prominent beneath; petiole 3–10 mm. long. Inflorescences of rather dense usually tomentose axillary spikes up to 10 cm. long usually appearing with the young leaves. Flowers (fig. 3/17, p. 14) yellow. Lower receptacle usually tomentose; upper receptacle broadly campanulate or cupuliform, 2·5 mm. long, 3·5 mm. in diameter, usually tomentose. Sepals deltate. Petals obcuneate, 1·1 mm. long, 0·6 mm. wide, ciliate at the apex. Stamen-filaments 6 mm. long; anthers 1·1 mm. long. Disk with pilose free margin 0·5–0·75 mm. wide. Style 2–3 mm. long, slightly expanded at the apex. Fruit (fig. 5/17, p. 16) subcircular or elliptic in outline, up to 3 cm. long, 3 cm. wide, usually crimson or pink, sometimes red-brown or tan, glutinous, nearly glabrous except for rather inconspicuous scales; apical peg absent; wings up to 13 mm. wide; stipe 2–10 mm. long. Cotyledons 2, arising below soil-level; petioles 4–4·5 cm. long. Scales (fig. 2/17, p. 13) glistening, whitish or yellowish or reddish, often obscured by the indumentum, ± 40–50(–70) μ in diameter, circular, with 8 primary radial walls only, cell walls very thin and obscured by glutinous secretions.

NOTE. Of the five subspecies (see Wickens in K.B. 26 : 37 (1971)), two occur in E. Africa.

subsp. **psidioides**; Exell in Mitt. Bot. Staats. Münch. 4 : 5 (1961); Stace in Mitt. Bot. Staats. Münch. 4 : 14 (1961) & in Bull. Brit. Mus. (Nat. Hist.) Bot. 4 : 27 (1965); Exell in Kirkia 7 : 197 (1970)

Leaf-lamina pubescent to densely pubescent or pilose on the reticulation beneath, sparsely pubescent to nearly glabrous (except for scales) on the areoles when mature. Young leaves densely grey or ferruginous pubescent. The size of the lamina varies in E. Africa from 9–20 cm. long and 8–12 cm. wide; the specimens with larger leaves have in the past been confused with subsp. *psilophyllum*.

KENYA. SE. Embu District, 17 Aug. 1965, *J.A. Robertson* 65 !
TANGANYIKA. Ngara District: Bushubi, Nuronzi, 5 Apr. 1961, *Tanner* 5922 !; Mpanda District: Mpanda–Tabora road, July 1956, *Procter* 469 !; Mbeya District: Chimala, 16 June 1952, *Mgaza* 505 !
DISTR. **K**4; **T**1, 3, 4, 7, 8; Zaire, Malawi, Mozambique, Zambia, Rhodesia, Botswana, Angola and South West Africa
HAB. *Brachystegia* and other deciduous woodland, often on rocky hill slopes; 900–1800 m.

SYN. *C. grandifolium* F. Hoffm., Beitr. Kenntn. Fl. Centr.-Ost-Afr.: 29 (1889); P.O.A. C: 289 (1895); Engl. & Diels in E.M. 3 : 39, t. 13/B (1899); T.T.C.L.: 136 (1949), pro parte. Type: Tanganyika, Mpanda District, Pa-Kabombue, *Boehm* 30a (B, holo. †, Z, iso. !)

C. grandifolium F. Hoffm. var. *retusa* F. Hoffm., Beitr. Kenntn. Fl. Centr.-Ost-
   Afr.: 30 (1889); Engl. & Diels in E.M. 3: 39 (1899). Type: Tanganyika,
   Tabora District, Igonda, *Boehm* 158a (B, holo. †, Z, iso.!)
C. grandifolium F. Hoffm. var. *eickii* Engl. & Diels in E.M. 3: 39 (1899). Type:
   Tanganyika, Lushoto District, Kwai, *Eick* (B, holo. †)
C. brachypetalum R. E. Fries in Wiss. Ergebn. Schwed. Rhod.-Kongo-Exped. 1:
   168, t. 1/4 & 13/4–6 (1914). Types: Zambia, Bangweulu, *Fries* 953 (S, syn., K,
   isosyn.!), 773, 953a & 965 (S, syn.)
C. psidioides Welw. subsp. *katangense* Duvign. in B.S.B.B. 88: 70 (1956), *nom.*
   non val. pub.

   subsp. **psilophyllum** *Wickens* in K.B. 26: 39 (1971). Type: Tanganyika, Ulanga
District, Ifakara, *Haerdi* 174/87 (K, holo.!, EA, iso.)

   Leaf-lamina 12–24 cm. long, 8–16 cm. wide, ferruginous pubescent along the sides of
the midrib and usually the lateral nerves as well, but glabrous on the areoles except for
the scales when mature. Young leaves glutinous and glabrescent.

TANGANYIKA.   Tabora District: Kaliuwa, 15 Sept. 1950, *Shabani* 13!; Ulanga District:
   Mufindi-Uzungwa-Ukwama Scarp, 29 Sept. 1951, *Gane* 11!; Songea District: near
   Tunduru District boundary, 5 June 1956, *Milne-Redhead & Taylor* 10499!
DISTR.   T1, ?3, 4–6, 8; Zaire, Zambia
HAB.   *Brachystegia* woodland on sandy and loamy soils; 350–1400 m.

SYN.   *C. vanderystii* De Wild., Ann. Mus. Congo, Bot., sér. 5, 3: 242 (1910). Types:
   Zaire, Léopoldville, Kisantu, *Vanderyst* (BR, syn.!, BM, fragment!) & *Gillet*
   430 bis (BR, syn.!)
   [*C. grandifolium* sensu T.T.C.L.: 136 (1949), pro parte, *non* F. Hoffm. sensu
   stricto]

NOTE.   This appears to be by far the commonest subspecies in East Africa. Although
   there have been sufficient gatherings of fruiting material with mature leaves very few
   flowering specimens have been gathered and more is required.
      A specimen of *Busse* 325 at Kew bears the manuscript name *C. anacardiifolium* Engl.
   The label bears no other information, but from the number it can be deduced that it
   probably comes from the Handeni District, near Kwa Sulanga. No other material
   has been seen from **T3**.

Sect. 8.   **Fusca** *Engl. & Diels* in E.M. 3: 76 (1899); Stace in J.L.S. 62: 160
(1969)

SYN.   Sect. *Coriifoliae* Engl. & Diels in E.M. 3: 75 (1899)

   Scandent shrubs and lianes. Flowers 4-merous. Upper receptacle elongate-campanulate, not divided into 2 regions. Petals oblanceolate to spathulate, sometimes minute. Stamens 8, 1-seriate. Disk without free margin.
Fruit, where known, 4-winged. Cotyledons unknown. Branches and
inflorescence-rhachides ± fuscous pubescent. Scales ± 30–90 μ in diameter,
circular in outline, convexly scalloped (or slightly so) at each marginal cell,
divided by (7–)8–16 primary radial walls alone.

Leaves twice as long as wide; tertiary nerves
   parallel; fruit 2–2·5 cm. long, subcircular in
   outline .   .   .   . 18. *C. fuscum*
Leaves thrice as long as wide; tertiary nerves not
   parallel; fruits not known .   .   .   . 19. *C. coriifolium*

   18. **C. fuscum** *Benth.* in Hook., Niger Fl.: 339 (1849); Laws. in F.T.A. 2:
426 (1871); Engl. & Diels in E.M. 3: 76, t. 23/A (1899); F.W.T.A., ed. 2, 1:
272 (1954); Liben in F.C.B., Combr.: 62 (1968). Types: Sierra Leone,
Freetown, *Vogel* 127 & Liberia, Grand Bassa, *Ansell* (both K, syn.!)

   Liane; branches at first fuscous pubescent, becoming glabrescent. Leaves
opposite; lamina coriaceous, ovate-elliptic to oblong-elliptic, up to 19 cm.
long and 9 cm. wide, apex acuminate, base obtuse to rounded, shortly

pubescent on the venation beneath, becoming glabrescent, minutely and rather inconspicuously lepidote; lateral nerves 6–9 pairs, tertiary nerves parallel; petiole 8–10 mm. long (–20 mm. *fide* F.C.B.), generally densely pubescent. Inflorescence of short rather crowded axillary spikes, appearing branched due to the suppression of the upper leaves, ± 2·5 cm. long on peduncles ± 1 cm. long; rhachis densely fuscous pubescent. Flowers (fig. 3/18, p. 14) cream. Lower receptacle densely tomentose; upper receptacle tubular-campanulate, 4 mm. long, 2 mm. in diameter, pubescent. Sepals deltate. Petals 1–2 mm. long, 1 mm. wide, glabrous. Stamen-filaments 6 mm. long, almost 1-seriate, at or very near the margin of the disk; anthers 0·9 mm. long. Disk-margin pilose, not free. Fruit (described from Nigerian material), fig. 5/18, p. 16, subcircular in outline, 2–2·5 cm. long, glabrous, glutinous, apex emarginate; apical peg absent; stipe 1–1·5 cm. long, pubescent. Cotyledons not known. Scales (fig. 2/18, p. 13) 35–50 μ in diameter, with (7–)8 primary radial walls.

UGANDA. Masaka District: Namalala Forest, *Fyffe* 51!
DISTR. **U**4; western Africa from Sierra Leone to Zaire and the Central African Republic
HAB. Forest; probably about 1300 m.

SYN. *C. sp.* sensu Burtt Davy, Check-lists Brit. Emp. 1, Uganda: 37 (1935), quoad specim. *Fyffe* 51

NOTE. More material of this very rare species is required.

19. **C. coriifolium** *Engl. & Diels* in E.M. 3: 75, t. 22/E (1899); T.T.C.L.: 134 (1949); Liben in B.J.B.B. 35: 179 (1965) & in F.C.B., Combr.: 62 (1968); Exell in Kirkia 7: 198 (1970). Type: Malawi, without locality, *Buchanan* 382 (B, holo. †, BM, K!, iso.)

Scandent evergreen shrub; branches at first fuscous pubescent, becoming glabrescent. Leaves opposite; lamina coriaceous, oblong to oblong-elliptic, up to 14 cm. long, 5·5 cm. wide, apex slightly and rather bluntly acuminate, base obtuse to rounded, glabrous or nearly so except for tufts of hairs in the axils of the nerves beneath and rather inconspicuously lepidote; lateral nerves 5–6 pairs, rather broadly spaced; petiole up to 9 mm. long, rather stout. Inflorescence of short crowded axillary or terminal spikes sometimes branched by suppression of the upper leaves, ± 1 cm. long on peduncles ± 6 mm. long; rhachis fuscous pubescent. Flowers cream. Lower receptacle tomentose; upper receptacle elongate-campanulate, 3 mm. long, 2 mm. in diameter. Sepals deltate. Petals spathulate, 2–2·2 mm. long, 0·9 mm. wide, glabrous. Stamen-filaments 4·5–5 mm. long, 1-seriate, at or very near the margin of the disk; anthers 0·9 mm. long. Disk-margin pilose, not free. Fruit unknown. Cotyledons unknown. Scales (fig. 2/19, p. 13) 40–60 μ in diameter, some with up to 16 primary walls.

TANGANYIKA. Lushoto District: Mangubu, 21 Oct. 1936, *Greenway* 4682!;* Morogoro District: Uluguru Mts., Kitundu, 27 Oct. 1934, *E. M. Bruce* 61!
DISTR. **T**3, 6; Malawi, Mozambique, Zaire (*fide* F.C.B.)
HAB. Lowland rain- and dry evergreen forest; 350–1400 m.

SYN. *C. laurifolium* Engl., P.O.A. C: 292 (1895), *non* Mart. (1839), *nom. illegit.* Type: as for species
C. leiophyllum Diels in E.J. 39: 506 (1907); T.T.C.L.: 141 (1949). Type: Tanganyika, Lushoto District, Amani, *Engler* 3440 (B, holo. †, BM, fragment!, EA, iso.)

NOTE. More material, especially fruiting, is required of this very rare species, which is apparently very close to 18, *C. fuscum.*

* A sheet at EA with otherwise the same data is numbered No. 4686.

Sect. 9. **Breviramea** *Engl. & Diels* in E.M. 3 : 61 (1899); Stace in J.L.S. 62 : 148 (1969)

Branching often divaricate. Flowers 4-merous. Upper receptacle campanulate to infundibuliform. Petals spathulate to broadly obovate or obovate or subcircular, glabrous. Stamens 2-seriate. Disk with a pilose free margin. Fruit 4-winged, oval to subcircular in outline, up to 3·5 cm. long, glabrous. Cotyledons 2, borne above soil-level. Scales (50–)75–120(–160) μ in diameter, irregular in outline, delimited usually by 8 primary radial walls with up to 8 tangential and up to 4 partial radial walls. Cells transparent or opaque.

Only one species in Flora area   .   .   .   .   20. *C. hereroense*

20. **C. hereroense** *Schinz* in Verh. Bot. Ver. Brand. 30 : 245 (1888); Engl. & Diels in E.M. 3 : 63 (1899); F.F.N.R.: 285 (1962); Exell in Kirkia 7 : 199 (1970); Wickens in K.B. 25 : 413 (1971). Type: South West Africa, Otjovazandu, *Schinz* 431 (Z, holo.)

Small tree up to 8(–12) m. high or more frequently (in E. Africa) a much branched coppicing shrub up to 5 m. high. Bark slightly fibrous, grey-brown to blackish; branchlets reddish brown to light grey, fibrous; young shoots tomentose to densely pubescent, later becoming glabrescent, or glabrous except for the ± contiguous rufous scales. Short horizontal spur shoots usually present. Leaves opposite or subopposite, usually borne along the length of the spur shoots but tending to cluster towards the apex; lamina coriaceous, narrowly elliptic to broadly obovate or subcircular (especially on sucker shoots), 3–6·5(–8) cm. long, 1·5–4(–4·5) cm. wide, apex slightly retuse to rounded or acute, mucronate, base cuneate to rounded, from densely tomentose to glabrous (except for the scales); scales silvery or golden and due to their wavy margins may present a rather scurfy appearance, contiguous to scattered, sometimes appearing impressed in subsp. *volkensii*; lateral nerves 3–9 pairs, subprominent to prominent beneath, visible above; petiole up to 5 mm. long. Inflorescence of short, sometimes rather congested spikes up to 3 cm. long, axillary, occasionally branched, often appearing on leafless branches (especially on spur shoots) in the axils of scars of fallen leaves; peduncles ± 0·7 cm. long. Flowers (fig. 3/20, p. 14) pale yellow to yellow, fragrant. Lower receptacle 1·5 mm. long; upper receptacle campanulate to shortly infundibuliform, 2–3 mm. long and wide, lepidote and otherwise densely pubescent to nearly glabrous. Sepals deltate. Petals spathulate to very broadly obovate or subcircular, 1·5–2·5 mm. long, 1–2·5 mm. wide, emarginate, clawed, glabrous. Stamen-filaments 2–4·5 mm. long; anthers 0·4–0·6 mm. long. Disk with a pilose margin free for 0·7–0·8 mm. Style 3–4 mm. long. Fruit (fig. 5/20, p. 16) oval to subcircular in outline, up to 3·5 cm. long and wide, often much smaller, glabrous except for the scales; apical peg absent or very short; stipe up to 11 mm. long. Cotyledons with petioles 10–11 mm. long; germination epigeal. Scales (fig. 2/20, p. 13) irregularly undulate in outline, (50–)75–100 μ in diameter, marginal cells 8–12, mostly retusely scalloped.

NOTE. This is a rather variable species and, as so often happens in this genus, with a reticulation of characters that makes the formation of infraspecific taxa rather unsatisfactory. Five infraspecific taxa can be recognized, but it is possible that with further gatherings intermediates will be found that will render the present system impracticable.

KEY TO INFRASPECIFIC VARIANTS

Scales (fig. 2/20a–d, p. 13) ± contiguous on lower
surface of leaf, golden, often mixed with dark
red scales (scales may be concealed by indumentum):

Fruits (fig. 5/20a, p. 16) 2 cm. long or usually
larger; leaves at least 2 cm. long (subsp.
hereroense):

Leaves glabrous or glabrescent . . . var. **hereroense**

Leaves densely pubescent, the hairs often
concealing the scales, reticulation prominent . . . . var. **villosissimum**

Fruits (fig. 5/20b, c) 1·5–2 cm. long; leaves
usually not more than 2 cm. long . . subsp. **grotei**

Scales (fig. 2/20e, f) not contiguous on lower
surface of leaf, golden or silvery, marginal
cells less scalloped and therefore scales
neater in outline (subsp. *volkensii*):

Scales impressed; leaves glabrous . . . var. **volkensii**

Scales not impressed; leaves pubescent, at least
on the midrib . . . . var. **parvifolium**

subsp. hereroense

Fruit 2 cm. long or usually larger and generally in Flora area densely rufous lepidote.
Leaf-blades at least 2 cm. long.

var. hereroense

Leaf-blades narrowly obovate to broadly obovate, in Flora area 2–8 cm. long, 1–4·5
cm. wide, glabrous or glabrescent. Fruit (fig. 5/20a, p. 16) subcircular in outline,
2–2·5 cm. long, densely rufous lepidote.

KENYA. Baringo District: Marigat, 30 Oct. 1964, *Leippert* 5255!; ?Kitui District:
Kimango (? = Kimangau), *R. M. Graham* 280 in *F.D.* 1689!; Kilifi District: Hadu,
Jan. 1937, *Dale* in *F. D.* 3666!
TANGANYIKA. Handeni District: Kangeti (? = Kangata), 2 Sept. 1955, *Watkins* 602A!;
Morogoro District: Morningside, June 1953, *Semsei* 1235! & Uluguru, 17 Mar. 1935,
E. M. Bruce 921!
DISTR. **K**3, 4, 7; **T**3, 6, 8; Malawi, Mozambique, Zambia, Rhodesia, Botswana, Angola,
South West Africa and South Africa

SYN. *C. usaramense* Engl., P.O.A. C: 290 (1895); Engl. & Diels in E.M. 3: 62, t. 19/C
(1899); T.T.C.L.: 139 (1949). Type: Tanganyika, Morogoro District, Ukami,
Stuhlmann 6745 (B, holo. †)

var. **villosissimum** *Engl. & Diels* in E.M. 3: 63, t. 19/G (1899). Type: South Africa
Transvaal, Makapansberge, *Rehmann* 5471 (Z, holo.)

Leaves with very prominent reticulation and densely pubescent.

TANGANYIKA. Mwanza District: Nyamikoma, 31 Mar. 1953, *Tanner* 1335!; Musoma
District: Serengeti Plains, Lake shore, Oct. 1932, *Rounce* 218!; Singida District:
near Sekenke, 2 Nov. 1960, *Richards* 13485!
DISTR. **T**1, 4, 5; Malawi, Mozambique, Zambia, Rhodesia, Botswana, South West and
South Africa

SYN. *C. villosissimum* (Engl. & Diels) Engl., V.E. 3 (2): 705 (1921)
C. transvaalense Schinz var. *villosissimum* (Engl. & Diels) Burtt Davy, Fl. Pl. &
Ferns Transv. 1: 246 (1926)

NOTE. More material is required from Tanganyika.

subsp. **grotei** (*Exell*) *Wickens* in K.B. 25: 415 (1971). Type: Tanganyika, Moshi,
Grote 5071 (EA, holo., BM, fragment!)

Leaves densely lepidote with ± contiguous scales, lamina usually not more than 2 cm.
long. Fruits (fig. 5/20b, c, p. 16) small, 1·5–2 cm. long.

UGANDA. Karamoja District: Cholol, Oct. 1956, *J. Wilson* 280 ! & Amudat, *Tweedie* 77 in *Bally* 3898 !; Mbale District: Greek River Camp, Jan. 1936, *Eggeling* 2503 !
KENYA. Northern Frontier Province: Dandu, 17 Mar. 1952, *Gillett* 12562 !; W. Suk District: Kapenguria, 28 July 1938, *Pole-Evans & Erens* 1533 !; Baringo District: Lake Hannington, W. side, 19 Jan. 1969, *Faden & Napper* 69/047 !
TANGANYIKA. Pare District: Muheza, Pangani [Ruvu] R., 2 Feb. 1930, *Greenway* 2107 ! & Himo–Kisangiro, 31 Jan. 1936, *Greenway* 4540 !; Lushoto District: Mkomazi, Oct. 1946, *Yussif bin Mohamedi* 4 !
DISTR. U1, 3; K1–3; T2, 3; Sudan

SYN. *C. grotei* Exell in J.B. 75: 165 (1937); T.T.C.L.: 139 (1949)
[*C. volkensii* sensu I.T.U., ed. 2: 89 (1952), *non* Engl.]

subsp. **volkensii** (*Engl.*) *Wickens* in K.B. 25: 415 (1971). Types: Kenya, Kwale/Kilifi District, Duruma, Maji ya Chumvi [Tschamtei], *Hildebrandt* 2337 (B, syn. †) & Tanganyika, Tanga District, Doda, *Holst* 2940 (B, syn. †, K, isosyn. !) & Pangani, *Volkens* 473 (B, syn. †, BM, isosyn. !)

Scales on leaf undersurface not contiguous, less scalloped than for subsp. *hereroense* and hence neater in outline.

var. **volkensii**

Scales (fig. 2/20e, p. 13) impressed. Leaf-blades up to 8 cm. long, glabrous. Fruit (fig. 5/20 d, e, p. 16) 1·5–2·5 cm. long, 1·5–2 cm. wide, densely reddish lepidote, otherwise glabrous.

UGANDA. Karamoja District: Lodoketemit Catchment, 22 Apr. 1959, *Kerfoot* 894 ! (atypical)
KENYA. Northern Frontier Province: Ijara, 30 Sept. 1957, *Greenway* 9252 !; Kwale District: near Taru, 11 Sept. 1953, *Drummond & Hemsley* 4266 !; Lamu District: Iwezo, 23 Feb. 1955, *Power* in E.A.H. 10994 !
TANGANYIKA. Lushoto District: Mkomazi, 30 Nov. 1935, *B. D. Burtt* 5338 !; Kilosa District: Kilosa–Morogoro road, 25 Oct. 1961, *Semsei* 3342 !; Iringa District: 90 km. N. of Iringa, 17 July 1956, *Milne-Redhead & Taylor* 11232 !
DISTR. U1; K1, 7; T3, 5–7; Somali Republic (S)

SYN. *C. volkensii* Engl. [in Abh. Preuss. Akad. Wiss. 1894: 18 (1894), *nom. nud.*], P.O.A. C: 290 (1895); Engl. & Diels in E.M. 3: 25, t. 7/A (1899); T.T.C.L.: 139 (1949); K.T.S.: 147 (1961)
C. greenwayi Exell in J.B. 75: 165 (1937); T.T.C.L.: 139 (1949). Type: Tanganyika, Lushoto District, Mshwamba, *Greenway* 2030 (BM, holo. !, EA, K !, iso.)

var. **parvifolium** (*Engl.*) *Wickens* in K.B. 25: 416 (1971). Type: Tanganyika, Mwanza District, Usambiro, *Stuhlmann* 852A (B, holo. †)

Scales (fig. 2/20f, p. 13) not impressed. Leaves pubescent, at least on the midrib. Fruit up to 2·5 cm. long, 2·1 cm. wide, densely reddish lepidote and pubescent.

KENYA. Northern Frontier Province: Moyale, 24 July 1952, *Gillett* 13620 !; Kitui District: near Mombasa, 14 Apr. 1958, *Trapnell* 2398 !; Teita District: Voi, Nov. 1955, *Ossent* 133 !
TANGANYIKA. Shinyanga District: Seseku (? = Seke), 10 June 1931, *B. D. Burtt* 2528 !; Dodoma District: 17 km. on Itigi–Chunya road, 16 Apr. 1964, *Greenway & Polhill* 11469 !; Mbeya District: Usangu, Mapogoro–Madibira, 12 Sept. 1936, *B. D. Burtt* 5957 !
DISTR. K1, 4, 7; T1, 2, 5–7; not known elsewhere

SYN. *C. parvifolium* Engl., P.O.A. C: 290 (1895); Engl. & Diels in E.M. 3: 62, t. 19/B (1899); T.T.C.L.: 139 (1949)
C. bruchhausenianum Engl. & Diels in N.B.G.B. 2: 189 (1898) & in E.M. 3: 63 (1899); T.T.C.L.: 139 (1949). Type: Tanganyika, Kilosa, *Brosig* 8 (B, holo. †, K, fragment !)

DISTR. (of species as a whole). U1, 3; K1–4, 7; T1–3, 5–8; Sudan and Somali Republic (S.), southwards to Lesotho and Natal and west to Angola
HAB. (of species as a whole). Wooded grassland and *Acacia*, *Commiphora* bushland, sometimes dominant; 50–2700 m.

Sect. 10. **Chionanthoida** *Engl. & Diels* in E.M. 3 : 77 (1899); Stace in J.L.S.
62 : 139 (1969)

Syn. Sect. *Meruensia* Engl. & Diels in E.M. 3 : 20 (1899)

Flowers 4-merous. Upper receptacle campanulate to very narrowly infundibuliform or tubular, sometimes expanding to a cupuliform apex. Petals narrowly obovate or obovate-spathulate or subcircular to (rarely) transversely elliptic, glabrous or ciliate. Stamens 2-seriate, variously inserted. Disk usually inconspicuous, without a free margin. Fruit 4-angled (aberrant 5- or 6-angled forms known), more rarely 4-winged, if winged then the body much larger than in other sections. Cotyledons not known. Scales (50–)75–150 μ in diameter, often appearing impressed, ± circular in outline, divided by numerous radial walls with tangential walls absent or very few; marginal cells 15–35, radially elongated.

Leaves glabrous or sparsely pubescent on the
   nerves:
  Receptacle lepidote, otherwise glabrous; fruits
    subsessile (unknown in *C. chionanthoides*):
    Receptacle campanulate; scales inconspicuous
      on mature leaves; petals ciliate .   .   21. *C. illairii*
    Receptacle infundibuliform; scales conspicuous
      on leaves:
      Leaf-lamina up to 8 cm. long; lateral nerves
        5–8 pairs; petals glabrous .   .   22. *C. butyrosum*
      Leaf-lamina up to 15 cm. long; lateral
        nerves 8–14 pairs; petals shortly
        ciliate .   .   .   23. *C. chionanthoides*
  Receptacle pilose and lepidote, infundibuliform;
    fruits with distinct stipe:
    Flowers in subglobose heads; leaves narrowly
      oblong or oblanceolate, margin distinctly
      undulate, with (10–)12–16 pairs of lateral
      nerves; stipe of fruit up to 3–5 mm.
      long .   .   .   .   24. *C. capituliflorum*
    Flowers spicate; leaves broadly elliptic, not
      distinctly undulate, with 7–10 pairs of
      lateral nerves; stipe of fruit up to 13 mm.
      long .   .   .   .   25. *C. xanthothyrsum*
Leaves distinctly hairy; receptacle pilose-lepidote:
  Receptacle infundibuliform; fruits subsessile .   26. *C. pisoniiflorum*
  Receptacle campanulate; fruits stipitate .   .   27. *C. exalatum*

21. **C. illairii** *Engl.* in Abh. Preuss. Akad. Wiss. 1894 : 15 (1894) & P.O.A. C : 289 (1895); Engl. & Diels in E.M. 3 : 80, t. 24/D (1899) [not fig. E as given on p. 80]; T.S.K., ed. 2 : 33 (1936); T.T.C.L. : 135 (1949); Exell in Kirkia 7 : 208 (1970). Type: Tanganyika, Tanga District, Moa [Muoa], *Holst* 3078 (B, holo. †, K, iso. !)*

Shrub to 4 m. high, or climber. Leaves opposite, subopposite or 3-verticillate; lamina coriaceous, oblong-elliptic, up to 14·5 cm. long, 6·5 cm. wide, apex acuminate, base rounded to subcordate, lepidote when young but appearing almost glabrous when mature with scales quite inconspicuous; lateral nerves 5–12 pairs, very prominent beneath; petiole up to 5 mm. long,

* Type not indicated in the original publication. *Holst* 3076 cited in error in P.O.A. C for 3078, as given in E.M.

stout, lepidote and white pubescent. Inflorescence of subcapitate axillary spikes usually ± 2 cm. long but may be up to 6 cm.; peduncles 1–1·5 cm. long, lepidote and rufous pubescent, stout and becoming very woody at the fruiting stage. Flowers yellow, fragrant. Lower receptacle 1·5 mm. long, lepidote and otherwise glabrous; upper receptacle campanulate, 1·5–2·5 mm. long, 2·2 mm. wide, lepidote and otherwise glabrous. Sepals scarcely developed. Petals obovate, 1·5 mm. long, 0·8 mm. wide, apex ciliate. Stamen-filaments 4–5 mm. long; anthers 0·8 mm. long. Disk rather inconspicuous. Style 4 mm. long. Fruit (fig. 5/21, p. 16) oblong or ovate in outline, 4-angled or very narrowly 4-winged, 1·3–2·5 cm. long, 1–1·5 cm. wide, glabrous or rarely rufous pubescent; apical peg scarcely developed; stipe very short or fruit subsessile. Scales (fig. 2/21, p. 13) as for the section.

KENYA. Northern Frontier Province: Galma Galla, 3 Sept. 1945, *J. Adamson* 141!; Kilifi District: Arabuko Forest, Jilori, Oct. 1965, *Tweedie* 3191!; Lamu District: Boni Forest, Mararani, 18 Sept. 1961, *Gillespie* 372!

TANGANYIKA. Tanga District: Pongwe–Maweni, Oct.-Nov. 1965, *Faulkner* 3685! & 3713!; Pangani District: Msubugwe, 26 Aug. 1955, *Tanner* 2111!; Uzaramo District: Dar es Salaam, Mar. 1868, *Kirk*!

ZANZIBAR. Zanzibar I., Unguja Ukuu, 5 Dec. 1930, *Greenway* 2666!

DISTR. **K**1, 7; **T**3, 6, 8; **Z**; Mozambique

HAB. A wide range of habitats from *Brachystegia* woodland and evergreen forest to coastal bushland; 30–800 m.

SYN. *C. meruense* Engl. [in Abh. Preuss. Akad. Wiss. 1894: 14 (1894), *nom. nud.*]*, P.O.A. C: 291 (1895); Engl. & Diels in E.M. 3: 20, t. 6/B (1899). Types: Tanganyika, Tanga District, Misoswe, *Holst* 2221 (B, syn. †, K, isosyn.!) & Handeni District, Kwa Mberue [Merue], *Fischer* 260 (B, syn. †, BM, fragment!)

C. hildebrandtii Engl., P.O.A. C: 289 (1895); Engl. & Diels in E.M. 3: 79, t. 24/E (1899) [not fig. D as stated on p. 79]; T.T.C.L.: 135 (1949); U.O.P.Z.: 207 (1949). Type: Tanganyika, Uzaramo District, Dar es Salaam, *Hildebrandt* 1248 (B, holo. †, BM, K!, iso.)

C. melchiorianum H. Winkler in F.R. 18: 123 (1922); T.T.C.L.: 136 (1949). Type: Tanganyika, Pare District, Makanya–Same, *Winkler* 3755 (BRSL, holo.)

C. schliebenii Exell & Mildbr. in N.B.G.B. 14: 106 (1938); T.T.C.L.: 136 (1949). Type: Tanganyika, Lindi, *Schlieben* 6480 (B, holo. †, BM, iso.!)

[*C. butyrosum* sensu K.T.S.: 143 (1961), *non* (Bertol. f.) Tul.]

NOTE. This is a rather variable species that may possibly be subdivided into two varieties when more correlated flowering and fruiting material becomes available. At present *C. illairii* can be distinguished from *C. hildebrandtii* only in the fruiting stage by having fruits oblong in outline with a retuse apex as opposed to ovate fruits with an acute apex, but even this character is unsatisfactory due to intermediate forms.

22. **C. butyrosum** (*Bertol. f.*) *Tul.* in Ann. Sci. Nat., sér. 4, 6: 87 (1856); P.O.A. C: 293 (1895); Engl. & Diels in E.M. 3: 81 (1899); Keay in K.B. 5: 342 (1951); Exell in Kirkia 7: 206 (1970). Type: Mozambique, Inhambane, *Fornasini* (BOLO, holo., BM, fragment!, P, iso.)

Scandent shrub or climber, or prostrate shrub or shrublet. Stem yellowish brown, pubescent; bark peeling to reveal reddish brown flaky surface. Leaves opposite; lamina papyraceous to coriaceous, oblong-elliptic to slightly obovate-oblong, up to 8 cm. long, 4·5 cm. wide, apex rounded and often apiculate, base subcordate, glabrous (except for scales) except for pubescence on the midrib and a few hairs on the lateral nerves beneath; scales golden, appearing impressed, widely scattered; lateral nerves 5–8 pairs, sparsely pubescent; petiole up to 5 mm. long, pubescent. Inflorescence of rather dense axillary spikes 1·5–3 cm. long on densely pubescent peduncles up to

* Exell, *loc. cit.* (1970), discusses the synonymy and considers *C. illairii* to be validly published by Engler in 1894 as the short diagnosis begins " ausgezeichnet durch . . . " while that for *C. meruense* is scarcely adequate.

4 cm. long. Flowers whitish to golden yellow, fragrant. Lower receptacle ± 1 mm. long, tomentose; upper receptacle infundibuliform, 3–7 mm. long, 2·5–3 mm. wide, pubescent when young to nearly glabrous when mature, often leaving a pubescent fringe to the calyx, lepidote. Sepals deltate, often pubescent along the margin. Petals obovate-spathulate, 1·5–2 mm. long, 0·9 mm. wide, glabrous. Stamen-filaments 4–5 mm. long, 2-seriately inserted near the mouth of the upper receptacle; anthers 0·4–0·8 mm. long. Disk inconspicuous. Fruit (fig. 5/22, p. 16) ellipsoid or obovoid-ellipsoid to sub-globose, 4(5)-angled or very narrowly 4(5)-winged, up to 3·7 cm. long, 2 cm. wide, usually tomentellous and lepidote (sometimes glabrous except for the scales), subsessile. Scales (fig. 2/22, p. 13) as for the section.

KENYA. Lamu District: Boni Forest, Maungi Pool, 25 Oct. 1957, *Greenway & Rawlins* 9428! & Mararani, 8 Nov. 1945, *J. Adamson* in *Bally* 5928! & 5929!
TANGANYIKA. Uzaramo District: Pugu Forest Reserve, June 1954, *Semsei* 1719!; Newala District: Mahuta, 12 Dec. 1942, *Gillman* 1064! & Mnalelo (? = Mnorera), 23 Mar. 1943, *Gillman* 1283!; Mikindani District: Nanguruwe, 12 Dec. 1942, *Gillman* 1085!
DISTR. K7; T6, 8; Mozambique
HAB. A rather rare species of dense thicket and riverine forest; 50–450 m.

SYN. *Sheadendron butyrosum* Bertol. f., Ill. Piante Mozamb. Dissert.: 12, t. 4/a, b (1850) in Mem. Accad. Sci. Ist. Bologna 2: 72 (1850)

NOTE. This species is closely related to 26, *C. pisoniiflorum* from which it differs in the less densely pubescent leaves and in the somewhat larger and more obovoid fruit. More material, especially fruiting, is required.
  Keay in K.B. 5: 342 (1951) refers to a specimen from Kenya, Kwale District, Mwachi, 1928, *R. M. Graham* in *F.D.* 1685 (BM, EA, K) as being *C. butyrosum*. This is a large leaved climber, lamina up to 11 cm. long and 6·5 cm. wide, with glabrous flowers, even in bud. More material of this variant is needed in order to determine its taxonomic status.

23. **C. chionanthoides** *Engl. & Diels* in E.M. 3: 79, t. 24/C (1899); T.T.C.L.: 135 (1949). Type: Tanganyika, Morogoro District, Ruvu R. near Tununguo, *Stuhlmann* 8971 (B, holo. †, K, fragment!)

Semi-scandent shrub or ? tree. Leaves opposite; lamina papyraceous, oblong-elliptic, up to 15 cm. long and 6 cm. wide (possibly longer and 7 cm. wide according to one leaf fragment), apex acuminate, base cordate, glabrous except for the scales, densely lepidote; scales golden-reddish, not contiguous; lateral nerves 8–14, rather prominent beneath; petiole ± 5 mm. long. Inflorescence axillary, simple or compound, ± 1·5 cm. long, hemispherical in outline; peduncle up to 1 cm. long, densely lepidote. Flowers yellowish, fragrant. Lower receptacle ± 2 mm. long; upper receptacle infundibuliform, 5–6 mm. long, 3 mm. wide, densely lepidote but otherwise glabrous. Sepals deltate. Petals elliptic, 1·5 mm. long, 0·8 mm. wide, clawed, shortly ciliate. Stamen-filaments ± 4·5 mm. long; anthers 0·5 mm. long. Disk inconspicuous. Style 8 mm. long. Fruit not seen. Scales (fig. 2/23, p. 13) as for other members of the section.

KENYA. Kwale District: Bome R., 15–16 Mar. 1902, *Kassner* 305! & 314! & Mwachi, *R. M. Graham* 0175
TANGANYIKA. Tanga District: Tanga–Korogwe, 8 Mar. 1966, *Leippert* 6415!; Pangani District: Bushiri Estate, 22 Dec. 1950, *Faulkner* 761!; Morogoro District: without precise locality, *Rounce* 495!
DISTR. K7; T3, 6; not known elsewhere
HAB. Forest margins; 50–170 m.

SYN. *C. stenanthum* Diels in E.J. 54: 342 (1917); T.T.C.L.: 135 (1949). Types: Tanganyika, Lushoto District, Amani, *Grote* in *Herb. Amani* 3429 & 3472 (both B, syn. †, BM, fragment!, EA, isosyn.!)

NOTE. More material, especially fruiting, is required of this rather rare species, also observations on habit and habitat.

24. **C. capituliflorum** *Schweinf.*, Reliq. Kotschy.: 33 (1868); Laws. in F.T.A. 2: 425 (1871); Engl. & Diels in E.M. 3: 78, t. 24/A (1899); Burtt Davy, Check-lists Brit. Emp. 1, Uganda: 35 (1935); F.P.S. 1: 203 (1950); Liben in F.C.B., Combr.: 53 (1968). Types: Sudan, Fazogli [Fesoglu], *Kotschy* 468 (W, syn., BM, K, isosyn.!) & 475 (W, syn., K, isosyn.!) & *Cienkowski* 108 & 146 (both W, syn.!)

Semi-scandent shrub ± 3 m. high to liane climbing to at least 16 m. Yellow-brown bark peeling rapidly to reveal light brown wood; young shoots sparsely to densely yellow-brown pubescent. Leaves opposite, subopposite or ternate; lamina membranous to coriaceous, oblong-oblanceolate, up to 23·5 cm. long, 5·5 cm. wide, apex acute to apiculate or acuminate, base cordate, margins undulate, glabrous (except for the scales) or pilose on the midrib beneath; scales golden reddish, scattered; lateral nerves 12–16 pairs, subprominent beneath; petiole up to 5 mm. long, yellow-brown pubescent at first, densely lepidote. Inflorescence axillary, simple or paniculate due to suppression of the leaves, spikes somewhat globose, ± 1·5 cm. in diameter; peduncle up to 1·5 cm. long, yellow-brown pubescent. Flowers greenish-white to yellow, odour not recorded. Lower receptacle 3·5 mm. long; upper receptacle infundibuliform, 5–6 mm. long, 2·5–3 mm. wide, densely to lightly pubescent, lepidote. Sepals deltate-subulate, 1·5 mm. long. Petals oblanceolate, 2 mm. long, 0·8 mm. wide, ciliate (glabrous according to F.C.B.). Stamen-filaments 5 mm. long; anthers 0·8 mm. long. Disk inconspicuous, with no free margin. Style 7 mm. long, shorter than the stamens. Fruit (fig. 5/24, p. 16) ovate in outline, ± 3–4 cm. long, 1·5–2 cm. wide, glabrous except for the scales; wings somewhat glossy, usually 4, sometimes 5 or 6, 3–4 mm. wide; apical peg absent; stipe up to 3–5 mm. long. Scales (fig. 2/24, p. 13) 100–150 μ in diameter, marginal cells not or only slightly scalloped; cells numerous, greatly radially elongated.

UGANDA. W. Nile District: Koka, 22 Mar. 1945, *Greenway & Eggeling* 7241!; Acholi/ Bunyoro District: Murchison Falls, Jan. 1952, *Leggat* 63!; Acholi District: Aswa [Oswa] R. on Gulu–Kitgum road, *Eggeling* 776!
KENYA. Turkana/W. Suk District: Turkwel R., Jan. 1957, *J. Wilson* 310!
DISTR. **U**1; **K**2; Zaire, Sudan and Ethiopia
HAB. Riverine forest; ± 800 m.

SYN. *C. undulato-marginatum* De Wild. & Exell in J.B. 67: 176 (1929); F.P.S. 1: 203 (1950). Type: Zaire, Orientale, Ituri R., Penge, *Bequaert* 2113 (BR, holo., BM, iso.!)
    C. lebrunii Exell in Rev. Zool. Bot. Afr. 21: 90 (1931). Type: Zaire, Equateur, Banzyville–Yakoma, *Lebrun* 2160 (BR, holo., BM, K, iso.!)

25. **C. xanthothyrsum** *Engl. & Diels* in E.J. 39: 507 (1907); T.T.C.L.: 141 (1949); Exell in Kirkia 7: 205 (1970). Type: Tanganyika, Dar es Salaam, *Holtz* 658 (B, syn. †, BM, fragment!) & Lindi District, Namgaru valley, *Busse* 2931 (B, syn. †, BM!, EA, isosyn.)

Semi-scandent evergreen shrub or liane. Leaves opposite; lamina membranous, elliptic to narrowly elliptic, up to 17·5 cm. long, 8 cm. wide, apex acute or acuminate, base usually rounded, sometimes subcuneate, glabrous except for the scales, rather densely lepidote on both surfaces; scales reddish golden, appearing deeply impressed, not contiguous; lateral nerves 7–10 pairs; petiole 5–15 mm. long. Inflorescence axillary, simple or paniculate due to the suppression of the leaves; spikes 3–5 cm. long; peduncle up to 2 cm. long, rufous pubescent. Flowers whitish to pale yellow, fragrant. Lower receptacle 2·5 mm. long, densely rufous lepidote; upper receptacle infundibuliform, 4·5 mm. long, 2·5 mm. wide, lepidote and densely pubescent. Sepals broadly deltate; petals broadly elliptic to subcircular, 1·5–2·5 mm. long, 1·5 mm. wide, conspicuously emarginate, glabrous. Stamen-filaments

5–6 mm. long; anthers 0·5 mm. long. Disk inconspicuous. Style 5 mm. long.
Fruit (fig. 5/25, p. 16) oblong-elliptic in outline, up to 5·5 cm. long and 3·5
cm. wide, lepidote and otherwise glabrous or nearly so; wings up to 10 mm.
wide and somewhat decurrent at the base; apical peg up to 1 mm. long; stipe
up to 1·3 cm. long. Scales (fig. 2/25, p. 13) as for the section.

TANGANYIKA. Lushoto District: Sigi–Derema [Nderema] road, 23 July 1953, *Drummond
   & Hemsley* 3420!; Morogoro District: Manyangu Forest Reserve, 24 Sept. 1959,
   Mgaza 328!; Uzaramo District: Kazimzumbwi Forest Reserve, Sept. 1964, *Shabani*
   in *Procter* 2759!
DISTR. T3, 6, 8; Mozambique
HAB. Secondary lowland rain-forest, dry evergreen forest and coastal bushland,
   seemingly rare; 200–800 m.

SYN. *C. stenanthoides* Mildbr. in N.B.G.B. 14: 105 (1938); T.T.C.L.: 141 (1949). Type:
   Tanganyika, Lindi District, Lake Lutamba, *Schlieben* 5212 (B, holo. †, BM,
   iso.!)

NOTE. This species can be readily distinguished from all other species of the section in
the Flora area by the broad-winged fruits.

26. **C. pisoniiflorum** (*Klotzsch*) *Engl.*, P.O.A. C: 293 (1895); Exell in
Kirkia 7: 207 (1970). Type: Mozambique, Manica e Sofala, Sena, *Peters*
(B, holo. †, K, fragment!)

Shrub up to 3–4 m. high, usually ± scandent, or a liane. Leaves opposite;
lamina papyraceous to coriaceous, elliptic or ovate-elliptic or narrowly
oblong-elliptic, up to 8(–9·5) cm. long and 4 cm. wide, apex shortly acuminate
or rounded, base rounded to cordate, glabrescent and shiny above, tomentose
(when young) to densely pubescent beneath even when mature, especially
on the nerves and reticulation, rather densely lepidote beneath, scales mostly
not quite contiguous and not appearing impressed although leaving depres-
sions (scales may be concealed by the indumentum); lateral nerves 4–7 pairs,
somewhat impressed above, prominent beneath; petiole 2–4 mm. long,
tomentose. Inflorescence of axillary spikes 2–3 cm. long; peduncles 1–2 cm.
long, tomentose. Flowers (fig. 3/26, p. 14) white or yellow, fragrant. Lower
receptacle 1 mm. long; upper receptacle infundibuliform, 3–5 mm. long,
pubescent and lepidote. Sepals broadly triangular. Petals narrowly obovate
to spathulate, 1·8–2 mm. long, 0·8–1·5 mm. wide, often emarginate. Stamen-
filaments 4·5–6 mm. long; anthers 0·4–0·5(–0·7) mm. long. Disk inconspicuous. Style 5 mm. long. Fruit (fig. 5/26, p. 16) narrowly ovoid-prismatic
to ovoid-prismatic, 2–3 cm. long, 1·2–1·7 cm. wide, apex sometimes acuminate,
4-angled or 4-ridged or very narrowly 4-winged, somewhat woody, tomentellous to glabrous (except for the scales) and lepidote (scales may be hidden
by the indumentum), dehiscing fairly readily along the angles, subsessile.
Scales (fig. 2/26, p. 13) with sometimes as few as 15–16 marginal cells,
otherwise as for the section.

TANGANYIKA. Pangani District: Bweni, Sept. 1955, *Semsei* 2268! & Bushiri, 3 Oct.
   1950, *Faulkner* 687!; Masasi District: Nangua, Nov. 1951, *Eggeling* 6382!
DISTR. T3, 8; Malawi, Mozambique
HAB. Riverine forest and thicket; 300–800 m.

SYN. *Sheadendron molle* Klotzsch in Peters, Reise Mossamb., Bot. 1: 76 (1861). Type:
   Mozambique, Manica e Sofala, Sena, *Peters* (B, holo. †)
   S. pisoniiflorum Klotzsch in Peters, Reise Mossamb., Bot. 1: 77, t. 14 (1861)
   [*Combretum pisoniaeflorum* in tab.]
   S. pisoniiflorum Klotzsch var. *brachystachyum* Klotzsch in Peters, Reise Mossamb.,
   Bot. 1: 77 (1861). Type: Mozambique, Sena, *Kirk* (B, holo. †)
   S. pisoniiflorum Klotzsch var. *macrostachyum* Klotzsch in Peters, Reise Mossamb.,
   Bot. 1: 77 (1861). Type: Mozambique, Sena, *Kirk* (B, holo. †)
   Combretum tetragonum Laws. in F.T.A. 2: 430 (1871); Engl. & Diels in E.M.
   3: 80, t. 24/F (1899), *non* Presl (1836), *nom. illegit.* Type: Mozambique,
   Manica e Sofala, Sena, *Kirk* (K, holo.!)

C. molle (Klotzsch) Engl. & Diels in E.M. 3: 80, excl. cit. tab. F (1899), *non* G. Don (1827), *nom. illegit.*

27. **C. exalatum** *Engl.*, P.O.A. C: 290 (1895); Engl. & Diels in E.M. 3: 81, t. 24/G (1899); T.T.C.L.: 136 (1949); K.T.S.: 144 (1961). Type: Tanganyika, Lushoto District, Mashewa [Mascheua], *Holst* 3559* (B, lecto. †, BM, K, fragments !, EA, iso. !)

Much-branched bushy scandent deciduous shrub up to 5 m. high. Young branches densely tomentose but peeling to expose a greyish purple surface. Leaves opposite; lamina papyraceous, narrowly oblong-elliptic to obovate, up to 13 cm. long and 4 cm. wide, apex rounded, sometimes retuse or even apiculate, base subcordate, revolute or folded inwards in time of drought, pubescent to glabrescent and lepidote above, pubescent to tomentose beneath, often concealing the dense but not contiguous silvery scales; lateral nerves 8–14 pairs; petiole up to 5 mm. long, stout, densely rufous pubescent. Inflorescence simple axillary spikes 2·3 cm. long; peduncles up to 2 cm. long, densely rufous pubescent. Flowers (fig. 3/27, p. 14) yellow-green to yellow, fragrance not recorded. Lower receptacle ± 1 mm. long, tomentose; upper receptacle campanulate, 2–3 mm. long, 2·5 mm. wide, tomentose to lightly pubescent, lepidote. Sepals deltate. Petals oblong-obovate, 1·5 mm. long, 0·8 mm. wide, glabrous. Stamen-filaments ± 4 mm. long, 2-seriately inserted near the base of the receptacle; anthers 0·5 mm. long. Disk lobed, not conspicuous. Fruit (fig. 5/27, p. 16) ovate-acute in outline, 2–3 cm. long, 1·2–1·5 cm. wide, shallowly winged (wings 1·5–7 mm. wide), glabrous; stipe 2–3 mm. long. Scales (fig. 2/27, p. 13) as for the section.

KENYA. Machakos District: Kiboko, 20 Feb. 1949, *Bogdan* 2371 !; Teita District: Voi, Oct.-Dec. 1955, *Ossent* in *E.A.H.* 10964 ! & Worssera Look-out, 15 Dec. 1966, *Greenway & Kanuri* 12764 !
TANGANYIKA. Pare District: Kisiwani, 2 Feb. 1936, *Greenway* 4571 !; Lushoto District: Lasa [Lassa] Mt., 30 Nov. 1935, *B. D. Burtt* 5328 !; Kilosa District: Kigowere–Lukandu [Lucando], 15 Feb. 1933, *B. D. Burtt* 4532 !
DISTR. **K**4, 7; **T**3, 6; Somali Republic
HAB. A common shrub of *Acacia, Commiphora* bushland; 300–1000 m.

SYN. *C. taitense* Engl. & Diels in E.M. 3: 80, t. 24/M (1899). Types: Kenya, Teita District, Ndi, *Hildebrandt* 2561 (B, syn. †, BM, K, isosyn. !) & probably Machakos District (below 1800 m.), *Scott Elliot* 6743 (B, syn. †, BM, K, isosyn. !)
      C. didymostachys Engl. & Diels in E.J. 39: 492 (1907). Type: Kenya, Kwale District, Samburu, *Kassner* 490 (B, holo. †, BM, K, iso.!)
      C. tavetense Diels in E.J. 39: 492 (1907); T.S.K., ed. 2: 33 (1936). Types: Kenya, Teita District, Taveta, *Engler* 1901 (B, syn. †, BM, fragment !) & Tanganyika, Pare District, Sengina–Simba, *Uhlig* 867 (B, syn. †, EA, isosyn. !) & Sadani–Kwa Ngoga [Kwagoge], *Engler* 1624 & 1653 (both B, syn. †)
      C. sennii Chiov., Fl. Somala 2: 211, fig. 128 (1932). Type: Somali Republic, R. Juba, Anole, *Senni* 54 (FI, holo. !, K, photo.)
      C. sp. sensu T.S.K., ed. 2: 34 (1936)

Subgenus CACOUCIA

(Aubl.) Exell & Stace in Bol. Soc. Brot., sér. 2, 40: 10 (1966)

Scales absent. Microscopic (and sometimes macroscopic) stalked glandular hairs always present. Flowers usually 5-merous (4-merous in sects. *Mucronata, Racemosa, Conniventia* and in *C. andradae* (sect. *Poivrea*)).

KEY TO SECTIONS

Flowers 4-merous:
  Petals glabrous or very minutely ciliolate:
    Upper receptacle (fig. 3/28, p. 14) cupuliform

---
* Engler in his original description included *Hildebrandt* 2561 but in his later monograph chose *Holst* 3559 as the lectotype and made *Hildebrandt* 2561 a syntype of his newly created *C. taitense,* now considered to be a synonym of *C. exalatum.*

to broadly campanulate, 2 × 2·5 mm.;
petals white; fruit up to 1·5 (–2) cm.
long; leaves pellucid-punctate   .   .   11. *Mucronata*

Upper receptacle (fig. 6/29, p. 50) campanulate
to tubular, 4–5 × 2–3 mm.; petals red;
fruit 2–2·5 cm. long; leaves not pellucid-
punctate .   .   .   .   .   12. *Conniventia*, p. 52

Petals pilose outside:
Inflorescence of terminal panicles with white
or coloured leaf-like bracts   .   .   13. *Racemosa*, p. 54
Inflorescence of lateral spikes, often in the
axils of fallen leaves; leaf-like bracts
absent   .   .   .   .   .   15. *Poivrea*, p. 57

Flowers 5-merous:
Petals glabrous:
Petals red, about as long as broad, ± connivent   12. *Conniventia*, p. 52
Petals white or pinkish, at least twice as long
as broad   .   .   .   .   .   15. *Poivrea*, p. 57

Petals pilose or pubescent externally:
Upper receptacle (fig. 6/33, p. 50) campanulate;
fruit sessile or nearly so   .   .   14. *Lasiopetala*, p. 56
Upper receptacle (fig. 6/35, 40, 43, p. 50)
divided into an upper infundibuliform or
broadly infundibuliform region and a
lower subglobose or cylindric region (with
a constriction between the upper and lower
receptacles in 43, *C. constrictum*); fruit
stipitate (except in *C. constrictum*) .   .   15. *Poivrea*, p. 57

Sect. 11. **Mucronata** *Engl. & Diels* in E.M. 3: 31 (1899)

Flowers 4-merous (5-merous ? aberrant flowers are also known, but not
from East Africa). Upper receptacle cupuliform to broadly campanulate,
1·5–2·5 mm. long. Petals white, transversely elliptic to obtriangular, margins
minutely ciliate. Stamens 8, 2-seriate. Disk glabrous with narrow free
margin. Leaves pellucid-punctate.

Only one species for the section .   .   .   .   28. *C. mucronatum*

28. **C. mucronatum** *Schumach.* in Schumach. & Thonn., Beskr. Guin. Pl.:
184 (1827) & in K. Danske Vid. Selsk. Nat. Math. Afhandl. 3: 204 (1828);*
Laws. in F.T.A. 2: 426 (1871); Engl. & Diels in E.M. 3: 31, t. 6/A (1899);
F.W.T.A., ed. 1, 1: 221 (1927). Type: Ghana [Danish Guinea], Akwapim
[Aquapim], *Thonning* (C, holo. (not traced), G-DC, iso.)

Scrambling shrub or climber; young branches rufous villous. Leaves
opposite; lamina obovate, up to 16 cm. long and 8 cm. wide, apex abruptly

* Beskr. Guin. Pl. is presumed to have been extracted in advance and given indepen-
dent pagination. The title page of the reprint is dated 1827 and must be accepted
(Stafleu, Taxonomic Lit.: 437 (1967)). *C. smeathmannii* was also published in the
Trans. Linn. Soc. of 1827. In the absence of more precise dating the two publications
are considered to be of equal status. The two species were first combined under *C.
mucronatum* by Lawson in F.T.A. 2: 426 (1871) and this name must take precedence over
the later reversal of the specific epithets (Exell in J.B. 67: 139 (1929)). Exell (personal
communication) has information from Christensen that no one in Denmark received a
copy of Beskr. Guin. Pl. before 1829; this cannot, however, be used as evidence of date of
publication.

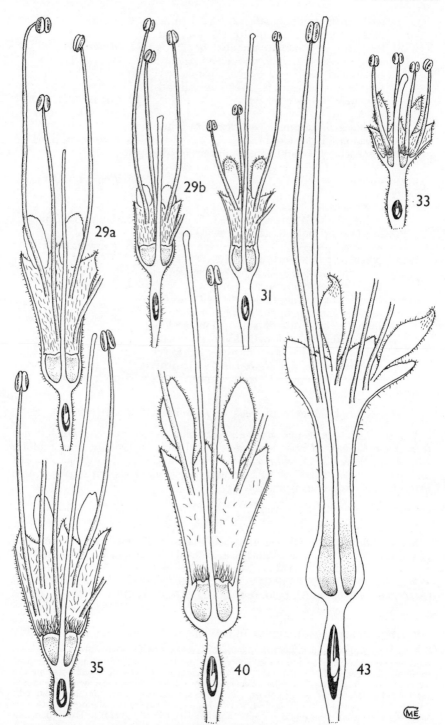

FIG. 6. Longitudinal sections, × 4, of flowers of *Combretum* species numbered as in text—**29**, *C. paniculatum* (a, subsp. *paniculatum*; b, subsp. *microphyllum*); **31**, *C. racemosum*; **33**, *C. obovatum*; **35**, *C. purpureiflorum*; **40**, *C. mossambicense*; **43**, *C. constrictum*. Drawn by Mrs. M. E. Church.

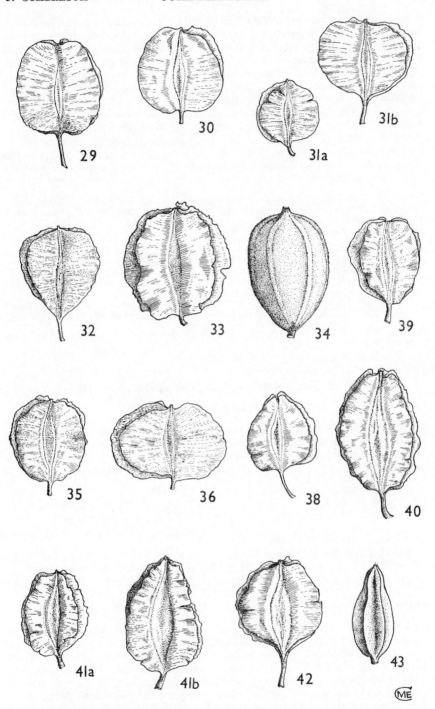

FIG. 7. Fruits, × 1, of *Combretum* species numbered as in text—**29**, *C. paniculatum*; **30**, *C. platypetalum*; **31 a, b,** *C. racemosum*; **32**, *C. cinereopetalum*; **33**, *C. obovatum*; **34**, *C. pentagonum*; **35**, *C. purpureiflorum*; **36**, *C. rhodanthum*; **38**, *C. aculeatum*; **39**, *C. longispicatum*; **40**, *C. mossambicense*; **41 a, b,** *C. goetzei*; **42**, *C. holstii*; **43**, *C. constrictum.* Drawn by Mrs. M. E. Church.

acuminate, base rounded to subcordate, lightly pubescent beneath, becoming glabrescent, pellucid-punctate; lateral nerves 7–11 pairs, prominent; petiole 4–8 mm. long, pubescent. Inflorescence of crowded axillary panicles up to 9 cm. long on peduncles 2–4 cm. long; rhachis rufous pubescent. Flowers (fig. 3/28, p. 14) white. Lower receptacle 2–3 mm. long; upper receptacle cupuliform, 1·5 mm. long, 2 mm. in diameter, rufous tomentose. Sepals shallowly deltate. Petals transversely elliptical to obtriangular, 2–3 mm. long and wide, margins minutely ciliate. Stamen-filaments 4 mm. long, 2-seriate; anthers 0·5 mm. long. Disk glabrous with narrow free margin. Fruit (fig. 5/28, p. 16, not yet known from East Africa) suborbicular to broadly obovate in outline, 1–1·8 cm. long, apex emarginate; wings papyraceous, straw-coloured, sometimes tinged with purple, minutely pubescent; apical peg short; stipe 1–2 mm. long.

UGANDA. Toro District: Bwamba, Kabango, 20 Jan. 1932, *Hazel* 152! & Sempaya, 22 Nov. 1935, *A. S. Thomas* 1518A! & Bwamba Forest, Oct. 1940, *Eggeling* 4053!
DISTR. U2; western Africa from Senegal southwards to Zaire
HAB. Secondary and riverine forest; 800 m.

SYN. *C. smeathmannii* G. Don in T.L.S. 15: 415, 424 (1827); Exell in J.B. 67: 139 (1929); F.W.T.A., ed. 2, 1: 272 (1954); Liben in F.C.B., Combr.: 59 (1968). Types: Sierra Leone, *Smeathmann* & *Afzelius* (both BM, syn.)
   C. sp. sensu Burtt Davy, Check-lists For. Trees & Shrubs Brit. Emp. 1, Uganda: 37 (1935), pro specim. *Hazel* 152!
   C. pellucidum Exell in Ann. & Mag. Nat. Hist. Lond., ser. 12, 7: 80 (1954). Type: Zaire, W. of Ruwenzori, Beni, *Mildbraed* 2298 (HBG, holo.)

NOTE. More material and information on the distribution of this species in Uganda is required.

Sect. 12. **Conniventia** *Engl. & Diels* in E.M. 3: 69 (1899)

SYN. Sect. *Parvula* Engl. & Diels in E.M. 3: 67 (1899)

Flowers 4(–5)-merous. Upper receptacle campanulate to tubular. Petals usually red, subcircular, exceeding the sepals, usually overlapping and ± connivent. Stamens 8(10), 2-seriate. Disk rather inconspicuous, without a free margin. Cotyledons 2, remaining below ground without unfolding.

Vigorous climber to scandent or ± prostrate
   shrub   .   .   .   .   .   .   29. *C. paniculatum*
Rhizomatous shrublet   .   .   .   .   30. *C. platypetalum*

29. **C. paniculatum** *Vent.*, Choix Pl., sub. t. 58 (1808); Laws. in F.T.A. 2: 425 (1871); Engl. & Diels in E.M. 3: 70, t. 21/C (1899), excl. var. *virgatum*; Burtt Davy, Check-lists Brit. Emp. 1, Uganda: 36 (1935); T.S.K., ed. 2: 32 (1936); T.T.C.L.: 134 (1949); U.O.P.Z.: 207 (1949); Exell in Kirkia 7: 212 (1970). Type: Senegal, *Roussillon* (P-JU, holo., IDC microfiche!)

A vigorous climber or scrambling or ± prostrate shrub; branchlets usually rufous tomentose at first, becoming glabrescent. Leaves opposite; lamina chartaceous, subcircular, oblong-elliptic, ovate-oblong or obovate-oblong, up to 18 cm. long and 9·5 cm. wide, apex rounded or acuminate, base obtuse to subcordate, either glabrous or nearly so (in East Africa) or greyish (drying silvery) or fulvous tomentose when young and becoming glabrescent; lateral nerves 4–8 pairs; petiole up to 3 cm. long, base often persistent and becoming spiny. Inflorescence of terminal or axillary panicles usually appearing before the leaves; rhachis usually fuscous or fulvous tomentose; bracts ± 2 mm. long, often transitional to foliage leaves. Flowers (fig. 6/29, p. 50)

red, 4- or 5-merous.* Lower receptacle up to 4(–5) mm. long, usually somewhat constricted above and below the ovary, densely fulvous or rufous
pubescent (more rarely silvery sericeous tomentose); upper receptacle
campanulate (sometimes almost cup-shaped in W. Africa), 2·5–6·5(–8) mm.
long, 2–4 mm. wide, (in E. Africa the 4-merous forms range from 3–6 mm.
long, the 5-merous from (4·5–)5–6(–8) mm. long), densely to lightly fulvous
or rufous or greyish or silvery pubescent, often becoming glabrescent (the
5-merous forms are invariably glabrescent). Sepals variable, from almost
inconspicuous to triangular. Petals red, subcircular to ovate, usually somewhat overlapping at first and ± connivent, ± 2·5 mm. long and wide,
glabrous. Stamen-filaments 7–18 mm. long, usually red; anthers 0·9 mm.
long, red or purplish. Disk inconspicuous, without a free margin. Fruit
(fig. 7/29, p. 51) 4- or 5-winged, subcircular to oblong-elliptic or broadly
elliptic in outline, 2–4(–5) cm. long, 1·5–4 cm. wide, body sparsely pubescent
to glabrous; wings thin; stipe 5–10 mm. long.

subsp. **paniculatum**

Leaf-lamina up to 18 cm. long, 9·5 cm. wide, usually 1·5 times as long as wide, usually
glabrous; lateral nerves 4–8 pairs. Lower receptacle (fig. 6/29a, p. 50) glabrous to
minutely puberulous; upper receptacle 3·5–6(–8) mm. long. Stamen-filaments 7–18 mm.
long. Fruit 2–4(–5) cm. long, 1·5–4 cm. wide.

UGANDA. Ankole District: Ruizi R., 7 Mar. 1951, *Jarrett* 27 !; Mbale District: Tororo,
    Oct. 1930, *Brasnett* in *F.H.* 39 !; Mengo District: [Kampala–]Entebbe road, Oct. 1931,
    Hansford in *Snowden* 2313 !
KENYA. Laikipia Plateau, *Routledge* !; Kiambu District: Thiririka Falls, 28 Jan. 1952,
    Kirrika 164 !; N. Kavirondo District: Kakamega, 15 Dec. 1951, *Trapnell* 2193 !
TANGANYIKA. Arusha District: 8 km. E. of Arusha, 29 Oct. 1959, *Greenway* 9595 !;
    Pangani District: Bushiri Estate, 26 Nov. 1950, *Faulkner* 736 !; Morogoro District:
    Turiani, 5 Sept. 1933, *B. D. Burtt* 4818 !
ZANZIBAR. Zanzibar I., Chumbuni, 23 Aug. 1931, *Vaughan* 1450 ! & 1451 ! & without
    precise locality, *Toms* 140 !
DISTR. U1–4; K3–5, 7; T1–8; Z; throughout tropical Africa from Senegal eastwards to
    Ethiopia and south to Angola, Rhodesia and Mozambique
HAB. Forest margins, in areas of woodland with higher rainfall and riverine forest;
    10–2000 m.

SYN. *C. abbreviatum* Engl., P.O.A. C: 292 (1895); Burtt Davy, Check-lists Brit. Emp. 1'
    Uganda: 35 (1935). Types: Zanzibar I., Kidoti, *Hildebrandt* 969 (B, syn. †,
    BM, isosyn. !) & Tanganyika, Lushoto District, Mazinde [Masinde], *Holst* 3876
    (B, syn. †)
    C. ramosissimum Engl. & Diels in E.M. 3: 72, t. 21/D (1899). Types: Sierra
    Leone, *Afzelius* (B, syn.†) & Fernando Po, *Mann* 203 (B, syn. †, K, isosyn. !)
    & Cameroun, Yaoundé, *Zenker & Staudt* 177 (B, syn. †, BM, isosyn.!)
    & *Zenker* 210 (B, syn. †) & Gabon, Mukungu, *Soyaux* 86 (B, syn. †, K, isosyn. !)
    C. buvumense Bak. f. in J.L.S. 37: 152 (1905); Burtt Davy, Check-lists Brit. Emp.
    1, Uganda: 35 (1935). Type: Uganda, Lake Victoria, Buvuma I., *Bagshawe*
    624 (BM, holo. !)
    C. unyorense Bagshawe & Bak. f. in J.B. 46: 5 (1908); Burtt Davy, Check-lists
    Brit. Emp. 1, Uganda: 36 (1935). Type: Uganda, Bunyoro [Unyoro] District,
    near Hoima, *Bagshawe* 1462 (BM, holo. !)
    C. sp. sensu Burtt Davy, Check-lists Brit. Emp. 1, Uganda: 37 (1935), pro
    specim. *Dawe* 96 ! & 463 ! & *Fyffe* 109/13 !

NOTE. The Kew specimen of *Greenway & Kanuri* 12209, from near Arusha, is unusual in
    having pubescent leaves, a character normally found in subsp. *microphyllum*, but
    the East African Herbarium specimen of this gathering is glabrous.

subsp. **microphyllum** (*Klotzsch*) *Wickens* in K.B. 26: 66 (1971). Type: Mozambique,
Manica e Sofala, vicinity of Rios de Sena and Tete, *Peters* (B, holo. †, BM, fragment !)

Leaf-lamina usually subcircular and up to 4·5 cm. long and 4 cm. wide, but sometimes
ovate-oblong to obovate-oblong and up to 11 cm. long and 6 cm. wide, usually greyish

* The pentamerous form is found in Ethiopia, along the Uganda border of Zaire, in
Uganda, Kenya and in Tanganyika on Kilimanjaro. Intermediate forms with a mixture
of tetramerous and pentamerous flowers or apparently tetramerous flowers producing
5-winged fruits can also be found in Tanganyika. The taxonomic significance of these
forms is not fully understood.

(drying silvery) or fulvous tomentose when young and retaining some indumentum but sometimes glabrescent; lateral nerves 5–6 pairs.   Lower receptacle (fig. 6/29b, p. 50) sericeous tomentose; upper receptacle 3–4·5 mm. long.   Stamen-filaments 7–13 mm. long.   Fruit similar to subsp. *paniculatum* but usually ± 2 cm. long and wide.

TANGANYIKA.   Lindi District: Lindi, *Gillman* 1101! & Ndanda, 7 Mar. 1966, *McCusker & Mwimmbi* 192! & Nachingwea, 28 Sept. 1952, *Anderson* 796!
DISTR.   T8; Malawi, Mozambique, Zambia, Rhodesia, Botswana and South Africa (Transvaal)
HAB.   Woodland, wooded grassland, thicket and riverine forest, in hotter and drier areas than subsp. *paniculatum*; 10–500 m.

SYN.   *C. microphyllum* Klotzsch in Peters, Reise Mossamb., Bot. 1: 74 (1861); Laws. in F.T.A. 2: 427 (1871); P.O.A. C: 292 (1895); Engl. & Diels in E.M. 3: 70, t. 21/A (1899); Exell in Kirkia 7: 213 (1970)
         C. lomuense Sim, For. Fl. Port. E. Afr.: 62, t. 61/B (1909).   Type: Mozambique, without precise locality, *Sim* 6393 (not located)

30. **C. platypetalum** *Laws.* in F.T.A. 2: 433 (1871); Engl. & Diels in E.M. 3: 68 (1899); Exell in Kirkia 7: 214 (1970).   Type: Angola, Huila, Mumpula–Humpata, *Welwitsch* 4356 (LISU, holo., BM!, COI, K!, P, iso.)

Shrublet usually ± 15–30 cm. high, but sometimes a shrub up to 3 m. (not in Flora area), with a thick woody rhizome, often flowering when leafless (not in Flora area); branchlets tomentose to glabrous.   Leaves opposite, sub-opposite or alternate or occasionally 3-verticillate; lamina 1–4 cm. long, 1·4–5 cm. wide, extremely variable in size and shape (from subcircular to very narrowly elliptic) and in indumentum (glabrous to densely tomentose). Inflorescence and flowers 4-merous, indistinguishable from those of *C. paniculatum*.   Fruit (fig. 7/30, p. 51) as in *C. paniculatum* but more elliptic and rarely subcircular, sometimes up to 5·5 cm. long and 3·5 cm. wide, often with a distinct apical peg 2–6 mm. long.

DISTR.   (of species as a whole).   T4, 7; Zaire, Malawi, Mozambique, Zambia, Rhodesia, Botswana, Angola and South West Africa

NOTE.   The species can be separated from *C. paniculatum* only by its habit.   Its taxonomic status requires intensive cytological and ecological studies, as do other species of rhizomatous shrublets and their related scandent taxa.
   Four subspecies are recognized by Exell in Bol. Soc. Brot., sér. 2, 42: 24 (1968), only one of which occurs in East Africa.

subsp. **oatesii** (*Rolfe*) *Exell* in Bol. Soc. Brot., sér. 2, 42: 25 (1968) & in Kirkia 7: 216 (1970).   Type: Rhodesia, Matabeleland, *Oates* (K, holo. !)

Shrublet with glabrous to sparsely pubescent leaves up to 4 times as long as broad. Not leafless when flowering; upper receptacle glabrous to sparsely pubescent.

TANGANYIKA.   Mpanda District: Kabungu, 10 Aug. 1948, *Semsei* 103! & Kungwe-Mahali Peninsula, Kalya, 23 Aug. 1959, *Harley* 9413!; Mbeya District: Mbozi Plateau, 28 Sept. 1936, *B. D. Burtt* 5969!
DISTR.   T4, 7; Zaire, Malawi, Mozambique, Zambia, Rhodesia
HAB.   *Brachystegia* woodland and *Hyparrhenia*, *Themeda* grassland, appearing approximately one month after burning (more information required); 750–2300 m.

SYN.   *C. oatesii* Rolfe in Oates, Matabeleland, ed. 2: 399, t. 10 (1889); Liben in F.C.B., Combr.: 38 (1968)
         C. turbinatum F. Hoffm. in Beitr. Kenntn. Fl. Centr.-Ost-Afr.: 28 (1889); P.O.A. C: 293 (1895); Engl. & Diels in E.M. 3: 68 (1899).   Type: Tanganyika, R. Ugalla, Boga, *Boehm* 43A (B, holo. †, Z, iso.)
         [*C. platypetalum* sensu T.T.C.L.: 135 (1949), *non* Laws. sensu stricto]

Sect. 13. **Racemosa** *Engl. & Diels* in E.M. 3: 82 (1899)

Inflorescences of terminal panicles with coloured or white leaf-like bracts. Flowers 4-merous.   Upper receptacle campanulate-infundibuliform.   Petals

oblong-elliptic to oblong-obovate, pilose externally.   Stamens 8, 2-seriate.
Disk inconspicuous, pilose, without a free margin.

Upper receptacle glabrous (or nearly so); fruit
    glabrous .   .   .   .   .   .   31. *C. racemosum*
Upper receptacle shortly pubescent; fruit lightly
    puberulous .   .   .   .   .   32. *C. cinereopetalum*

31. **C. racemosum** *P. Beauv.*, Fl. Owar. 2: 90, t. 118 (1820); Laws. in F.T.A.
2: 424 (1871), excl. syn. *C. macrocarpum* P. Beauv.; P.O.A. C: 292 (1895);
Engl. & Diels in E.M. 3: 82, t. 23/C (1899), excl. syn. *C. macrocarpum*;
Bagshawe & Bak. f. in J.B. 46: 6 (1908); Burtt Davy, Check-lists Brit. Emp.
1, Uganda: 36 (1935). Type: Nigeria, Benin, *Palisot de Beauvois* (P, holo.)

A liane or sometimes a scandent shrub; branchlets lightly villous at first,
becoming glabrescent. Leaves opposite or subopposite, sometimes ternate;
lamina oblong-ovate to ovate, up to 16 cm. long and 6 cm. wide, apex shortly
acuminate, base rounded or subcordate, lightly pilose at first, becoming
glabrescent; lateral nerves 6–10 pairs; midrib prominent beneath; petiole up
to 5 mm. long. Inflorescence of terminal and axillary contracted racemes;
rhachis rufous pubescent; floral leaves up to 6 cm. long and 2·5 cm. wide,
similar to foliage leaves but whitish or pink. Flowers (fig. 6/31, p. 50)
crimson, 4-merous. Lower receptacle 3–4·5 mm. long, somewhat constricted
above and below the ovary, usually lightly pilose; upper receptacle campanulate, 5–6 mm. long, 3 mm. wide, glabrous or nearly so. Sepals shortly triangular, acuminate, 0·5 mm. long. Petals oblong-elliptic, 2–3 mm. long,
0·1–2 mm. wide, densely fuscous pubescent externally. Stamen-filaments
12 mm. long; anthers 0·5 mm. long. Disk inconspicuous, pilose, without a
free margin. Fruit (fig. 7/31, p. 51) obtriangular to obovate in outline,
1·8–2·5 cm. long, 1·7–2·5 cm. wide; wings membranous, glabrous; apical peg
absent or minute; stipe 3 mm. long.

UGANDA. Bunyoro District: Budongo Forest, Jan. 1941, *Purseglove* 1093! & Masindi–
    Butiaba, 7 Jan. 1953, *Verdcourt* 871!; Mubende District: without precise locality,
    Feb. 1930, *Brasnett* 1!
DISTR. U2, 4; western Africa from Senegal southwards to Angola and east to southern
    Sudan
HAB. Margins of rain-forest, riverine forest and in secondary bushland; 1100–2200 m.

NOTE. Sometimes grown as an ornamental, known as " False Bougainvillea ", e.g.
    Kenya, Fort Ternan, *Gardner* in F.D. 3719! and Trans-Nzoia District, Kapretwa,
    Tweedie 3714!

32. **C. cinereopetalum** *Engl. & Diels* in E.M. 3: 84, t. 23/E (1899); Burtt
Davy, Check-lists Brit. Emp. 1, Uganda: 35 (1935); Liben in F.C.B., Combr.:
29 (1968). Types: Cameroun, Buea, *Lehmbach* 136 (B, syn. †) & Yaoundé,
Zenker 740 (B, syn. †, K, isosyn.!) & Angola, Bango Aquitamba–Bumba,
Welwitsch 4353 (B, syn. †, BM!, COI, K!, LISU, P, isosyn.) & a number of
syntypes from Zaire

A liane or sometimes a scandent shrub; branchlets densely villous, becoming glabrescent. Leaves opposite or subopposite; lamina elliptic or broadly
elliptic to oblong, 5–8 cm. long, 3–4 cm. wide, apex shortly acuminate, base
truncate to subcordate, lightly pilose at first, becoming glabrescent except
on the nerves; lateral nerves 4–8 pairs; petiole up to 5 mm. long, sometimes
persisting and forming a spine. Inflorescence of terminal and axillary contracted racemes; rhachis densely pubescent; floral leaves resembling small
foliage leaves, but whitish and early deciduous. Flowers crimson, 4-merous.
Lower receptacle 3–5 mm. long, somewhat constricted above and below the

ovary, densely pilose; upper receptacle campanulate, 5–7 mm. long, densely pubescent. Sepals triangular, 1 mm. long. Petals obovate, 2·5–3 mm. long (3·5–4·5 mm. long *fide* F.C.B.), 1·5 mm. wide, densely fuscous pubescent externally. Stamen-filaments 12 mm. long; anthers 0·5 mm. long. Disk inconspicuous, pilose, without a free margin. Fruit (fig. 7/32, p. 51) obtriangular or obovate in outline, 2·2–2·5 cm. long, 2–2·3 cm. wide; wings membranous, lightly puberulous; apical peg minute; stipe 3 mm. long.

UGANDA. Mengo District: Mukono, Oct. 1914 (?1913), *Dummer* 283! & Kijude, Nov. 1915, *Dummer* 2674! & Nambigirwa, Mar. 1923, *Maitland* 616! & Entebbe, Jan. 1932, *Eggeling* 167!
DISTR. U4; Cameroun, Central African Republic, Zaire and Angola
HAB. Forest margins; 1050–1400 m.

SYN. *C. mittuense* Engl. & Diels in E.M. 3: 83, t. 23/D (1899). Type: Sudan, Bahr el Ghazal, R. Roah–Kuddu, *Schweinfurth* 2774 (B, holo. †, K, iso.!)
 C. cabrae De Wild. & Th. Dur. in B.S.B.B. 39: 100 (1901). Type: Zaire, Bas-Congo, *Cabra* 75 (BR, holo.)
 C. sp. sensu Burtt Davy, Check-lists Brit. Emp. 1, Uganda: 37 (1935), pro specim. *Maitland* 616!

NOTE. This is a rare species in Uganda, and apparently no recent gatherings have been made. Good fruiting and vegetative material has not been well collected from any part of its distribution range.

Sect. 14. **Lasiopetala** *Engl. & Diels* in E.M. 3: 65 (1899)

Flowers 5-merous. Upper receptacle campanulate. Petals oblong to ovate or subobovate, exceeding the sepals, hairy externally. Stamens 10, 2-seriate. Disk with scarcely free margin, glabrous or hairy.

Upper receptacle olive-green to olive-brown;
    fruit 5-winged, 3–3·5 cm. long   .   .   .   33. *C. obovatum*
Upper receptacle reddish brown; fruit 5-angled,
    2–4·5 cm. long   .   .   .   .   .   34. *C. pentagonum*

33. **C. obovatum** *F. Hoffm.*, Beitr. Kenntn. Fl. Centr.-Ost-Afr.: 28 (1889); P.O.A. C: 291 (1895); Engl. & Diels in E.M. 3: 66, t. 20/C (1899); T.T.C.L.: 134 (1949); Exell in Kirkia 7: 210 (1970). Type: Tanganyika, Tabora District, Igonda [Gonda], *Boehm* 140A (B, holo. †)

Usually a coppicing shrub, semi-scandent or scandent, sometimes a liane, possibly (*fide Procter* 1665) a small tree. Leaves opposite or 3-verticillate, often white or whitish and bract-like below the inflorescences; lamina obovate, broadly obovate-oblong or obovate-elliptic, up to 7(–8) cm. long and 3–4 cm. wide, apex usually rounded and occasionally rather bluntly acuminate and often mucronate, base usually rounded to subcordate, tomentose when young and remaining densely pubescent or pilose beneath when mature at least on the nerves and reticulation; lateral nerves 5–8 pairs; petiole 5–12 mm. long, sometimes persisting and forming a curved spine. Inflorescences of congested or subcapitate axillary spikes 2–4(–6) cm. long or short panicles by reduction of the leaves to leaf-like bracts. Flowers (fig. 6/33, p. 50) 5-merous, white, cream or pink, fragrant. Lower receptacle 2 mm. long, densely pubescent; upper receptacle broadly campanulate, 2–3 mm. long, 3 mm. in diameter, tomentose to densely pubescent. Sepals ovate-triangular, 1·5 mm. long. Petals narrowly oblong or obovate-oblong or spathulate, 2 mm. long, 0·6–0·8 mm. wide, hairy externally. Stamen-filaments 6–7 mm. long; anthers 1 mm. long. Disk near the base of the receptacle; margin lobed, thick, glabrous, scarcely free. Fruit (fig. 7/33, p. 51) 5(–6)-winged, obovate-oblong to transversely elliptic in outline, 3–3·5 cm. long, 2·5–3 cm. wide, subsessile, pale brown tomentose when young and remaining so on the body; apical peg absent or very short. Cotyledons 3, arising at or below soil-level and unfolding spirally; petioles 7–13 mm. long.

Tanganyika. Mwanza District: Missungwi, Budutu, 18 Oct. 1951, *Tanner* 3315!; Tabora District: Isimbira, 25 Oct. 1960, *Richards* 13392!; Mbeya District: Chimala, 10 Jan. 1963, *Napper* 1688!
Distr. T1, 2, 4, 5, 7; Mozambique, Zambia and Rhodesia
Hab. Coppicing shrub in fallow lands, a scandent shrub of deciduous bushland and a vigorous liane in riverine forest; 840–1500 m.

34. **C. pentagonum** *Laws.* in F.T.A. 2: 424 (1871); Engl. & Diels in E.M. 3: 102 (1899); T.T.C.L.: 134 (1949); Exell in Kirkia 7: 211 (1970). Type: Mozambique, Rovuma R., *Meller* (K, holo.!)

Woody climber; branchlets fulvous tomentose. Leaves opposite or sub-opposite, sometimes 3-verticillate; lamina papyraceous, elliptic to obovate-elliptic, up to 16(–20) cm. long and 8(–11·5) cm. wide, apex acuminate, base subcordate, almost glabrous above except for the lightly pubescent midrib and lateral nerves, densely fulvous pubescent to fulvous tomentose to almost glabrous beneath except for the nerves; lateral nerves 7–16 pairs; petiole up to 1·5 cm. long, the lower portion of which may form a spine. Inflorescence of subcapitate spikes or racemes 3–6 cm. long including the peduncle; rhachis fulvous pubescent. Flowers sessile or sometimes very shortly pedicellate, cream to pink, maroon or crimson, fragrant. Lower receptacle 4·5–5·5 mm. long, constricted above the ovary, sericeous; upper receptacle campanulate, 2–3 mm. long, 3–3·5 mm. in diameter, appressed pubescent. Sepals triangular, up to 1 mm. long. Petals ovate to subobovate, 1·5–2·5 mm. long, 0·9–1·1 mm. wide, hairy externally, especially on the margins. Stamen-filaments 3·5–6 mm. long; anthers 1 mm. long. Disk 2 mm. in diameter, with a pilose scarcely free margin. Style exserted before the stamens. Fruit (fig. 7/34, p. 51) 5-angled, narrowly obovate in outline, 2–4·5 cm. long, 1·2–2·5 cm. in diameter, apex slightly acuminate, subsessile or very shortly stipitate, glabrous.

Kenya. Kilifi District: Ribe, May 1880, *Wakefield*! & Takaunga, *R. M. Graham* D.424 in *F.D.* 1851! & Kibarani, 22 Feb. 1946, *Jeffery* 475!
Tanganyika. Pare District: Kisiwani, 27 Jan. 1937, *Greenway* 4874!; Tanga District: Magunga Estate, 3 Jan. 1953, *Faulkner* 1110!; Morogoro District: Kisaki, 6 Dec. 1933, *B. D. Burtt* 4993!
Distr. K7; T3, 6–8; Malawi, Mozambique and Zambia
Hab. Various types of forest, secondary associations and thicket; 20–900 m.

Syn. *C. wakefieldii* Engl., P.O.A. C: 291 (1895); Engl. & Diels in E.M. 3: 65, t. 20/B (1899); T.S.K., ed. 2: 33 (1936). Type: Kenya, *Wakefield* (B, holo. †)*
C. lasiopetalum Engl. & Diels in E.M. 3: 65, t. 20/A (1899). Types: Tanganyika, Morogoro District, Tununguo, *Stuhlmann* 8966 & Uzaramo District, Mbagala, *Stuhlmann* 9234 & Kazimzumbwi [Kasi], *Goetze* 17 (all B, syn. †)

Sect. 15. **Poivrea** (*Juss.*) *G. Don*, Gen. Syst. 2: 665 (1832)

Syn. *Poivrea* Juss., Gen. Pl.: 320 (1789)
Sect. *Grandiflora* Engl. & Diels in E.M. 3: 88 (1899)
Sect. *Trichopetala* Engl. & Diels in E.M. 3: 92 (1899)
Sect. *Malegassica* Engl. & Diels in E.M. 3: 110 (1899)

Flowers usually 5-merous, sometimes 4-merous. Upper receptacle cylindric to infundibuliform or elongate campanulate, sometimes constricted above the disk. Petals oblong-ovate to narrowly elliptic to spathulate,

* This may be the same gathering as the *Wakefield* specimen from Kilifi District, Ribe, 1880 (K!).

glabrous or pubescent externally.   Stamens (8-)10, 2-seriate.   Disk without a free margin.

Petals glabrous or almost so:
 Inflorescence secund spikes; fruit-stipe up to
  2 mm. long; leaves obovate . . . 35. *C. purpureiflorum*
 Inflorescence not secund; fruit-stipe 5–10 mm.
  long; leaves ovate-elliptic to elliptic-obovate 36. *C. rhodanthum*
Petals pilose or pubescent externally:
 Flowers and fruits 4-merous, densely hairy,
  not conspicuously glandular; fruit 3·5–5
  cm. long, velutinous . . . . 37. *C. andradae*
 Flowers and fruits normally 5-merous:
  Upper receptacle only slightly constricted
   between the upper and lower regions;
   margin of disk pilose; fruits winged:
   The upper receptacle densely hairy and/or
    with dense macroscopic glandular hairs;
    fruit pubescent or glutinous:
    Upper receptacle pubescent to silky
     tomentose, macroscopic glandular
     hairs absent or sparse; fruit pubes-
     cent, not glutinous:
     Upper receptacle 3–6 mm. long; leaves
      up to 7 cm. long and 5 cm. wide,
      usually smaller . . . . 38. *C. aculeatum*
     Upper receptacle 6–11 mm. long; leaves
      usually larger:
      Inflorescence a secund spike up to 22
       cm. long; upper receptacle 9–11
       mm. long, densely covered with
       long silky appressed hairs; fruit-
       stipe 4–5 mm. long . . 39. *C. longispicatum*
      Inflorescence not secund, up to 8 cm.
       long; upper receptacle 6–9 mm.
       long, with short mostly spreading
       hairs; fruit-stipe 4–10 mm.
       long . . . . . 40. *C. mossambicense*
    Upper receptacle 6–10 mm. long, with
     dense macroscopic glandular hairs;
     fruit glutinous . . . . 41. *C. goetzei*
   The upper receptacle usually glabrous except
    for occasional macroscopic glandular
    hairs, rarely sparsely hairy; fruit
    glabrous and shiny . . . 42. *C. holstii*
  Upper receptacle (fig. 6/43, p. 50) constricted
   between the upper and lower parts for a
   length of 5–6 mm.; margin of disk glab-
   rous; fruits 5-angled, ± 2·5 cm. long . 43. *C. constrictum*

35. **C. purpureiflorum** *Engl.*, P.O.A. C: 292 (1895); Engl. & Diels in E.M. 3: 90, t. 26/D (1899); T.T.C.L.: 133 (1949).   Type: Tanganyika, Tabora, *Stuhlmann* 570 (B, holo. †)

A multistemmed scandent shrub up to 7–8 m., climbing chiefly by aid of its petiolar spines, but also twining.   Leaves opposite, subopposite or alternate, sometimes on dwarf shoots in the axils of the petiolar spines; lamina obovate, up to 10 cm. long and 6 cm. wide, apex obtuse or slightly acuminate, base cuneate to almost subcordate, sparsely hairy, especially along the nerves; lateral nerves 5–7 pairs, rather prominent beneath; petiole up to 11 mm. long, later forming a slightly recurved spine. Inflorescence generally appearing before the leaves, usually with solitary secund spikes up to 7 cm. long in the axils of the old leaves. Flowers (fig. 6/35, p. 50) bright red, shortly pedicellate. Lower receptacle 3–4 mm. long, pubescent; upper receptacle infundibuliform, slightly constricted about the middle, 7–8 mm. long, 3 mm. in diameter, pubescent. Sepals triangular-subulate, 1·5–2 mm. long. Petals oblong-ovate, 3 mm. long, 1 mm. wide, glabrous or almost so. Stamen-filaments ± 20 mm. long, biseriate; anthers ± 1 mm. long, bright red. Disk-margin not free, pilose. Fruit (fig. 7/35, p. 51) 5-winged, oblong-ovate in outline, 2·5–3·5 cm. long, 2–2·5 cm. wide, glabrous; apical peg absent; stipe ± 1–2 mm. long.

TANGANYIKA. Mwanza District: near Kikonko Ferry, 17 July 1960, *Verdcourt* 2882!; Tabora District: Tumbi, 27 July 1954, *Joseph* 4016!; Dodoma, 13 Nov. 1958, *Napper* 860!
DISTR. T1, 4–8; not known elsewhere
HAB. Common in deciduous woodland, bushland and wooded grassland, in seasonally inundated valleys and on hill slopes; 840–1800 m.

36. **C. rhodanthum** *Engl. & Diels* in E.M. 3 : 92 (1899); F.W.T.A., ed. 2, 1 : 270, 274 (1954); Liben in F.C.B., Combr.: 32 (1968). Type: Sierra Leone, near Mafari, *Scott Elliot* 4438 (B, holo. †, BM, iso. !)*

A scandent shrub or liane; stem lightly rufous pubescent. Leaves opposite or sometimes 3-verticillate; lamina ovate-elliptic to elliptic-obovate, up to 21 cm. long and 11 cm. wide, apex acuminate, base cordate, glabrous when mature except for pilose domatia in the nerve-axils and sparsely hispid midrib and nerves beneath; lateral nerves 6–10 pairs; petiole up to 1 cm. long, stout and leaving a prominent leaf-scar. Inflorescence of elongated false racemes formed by the suppression of the leaves, 10(–17) cm. long; flowers pink or crimson, on very short ultimate branches of the false raceme, rather crowded in the upper part; peduncle and flowers without or with a very few long gland-tipped hairs, but short gland-tipped hairs sometimes present. Lower receptacle 5–7 mm. long, lightly pubescent; upper receptacle infundibuliform, 4–6(–9) mm. long, 3(–5) mm. in diameter, lightly pubescent. Sepals triangular-subulate, ± 1 mm. long. Petals oblong-elliptic, 6–8(–13) mm. long, (1–)2(–3) mm. wide, glabrous. Stamen-filaments 14–17(–22) mm. long; anthers 0·6 mm. long. Disk-margin not free, pilose. Fruit (fig. 7/36, p. 51) 5-winged, elliptic to broadly circular-elliptic in outline, 2·2–3 cm. long and wide, lightly pubescent; apical peg very short; stipe up to 10 mm. long.

UGANDA. Bunyoro District: Budongo Forest, Feb. 1935, *Eggeling* 1602!; Toro District: near Kabango, 20 Jan. 1932, *Hazel* 150!; Mengo District: Kijude, Dec. 1915, *Dummer* 2681!
DISTR. U2, 4; Sierra Leone, Liberia, Ivory Coast, Ghana, Cameroun, Zaire and Sudan
HAB. Rain-forest and secondary associations; 1050–1200 m.

SYN. [*C. trichopetalum* sensu Burtt Davy, Check-lists Brit. Emp. 1, Uganda: 36 (1935), *non* Engl.]
C. sp. sensu Burtt Davy, Check-lists Brit. Emp. 1, Uganda: 37 (1935), pro specim. *Hazel* 150!

* Engler & Diels erroneously cite the type as *Scott Elliot* 4638. An unpublished check-list of the Scott Elliot collection (the cancelled proof copy of part of a paper intended for the J.L.S. 30, in the Kew Library) gives on p. 140, No. 4638 as *Macaranga heterophylla* (Muell. Arg.) Muell. Arg. and on p. 119, Nos. 4438 and 4443 from Mafari as *Combretum sp.*

37. C. andradae *Exell & Garcia* in Contr. Conhec. Fl. Moçamb. 2 : 132, t. 11 (1954) [figs. C & D should be × 1, not × 2] ; Exell in Bol. Soc. Brot., sér. 2, 42 : 6 (1968) & Kirkia 7 : 218 (1970). Type : Mozambique, Niassa, Amaramba, *Andrada* 1436 (BM, holo.!, COI, LISC, iso.)

Woody climber or shrub ± 2 m. high ; branchlets at first tomentose but soon glabrescent. Leaves subopposite ; lamina membranous (at least when young), broadly elliptic, up to 19 cm. long and 12·5 cm. wide (*fide* Exell—all 3 sheets at Kew are leafless), densely minutely glandular above, tomentose to densely pilose beneath, apex rounded or acuminate or apiculate, base subcordate to cordate ; lateral nerves 7–10 pairs ; petiole up to 8 mm. long, tomentose. Inflorescence of lateral spikes 4–7 cm. long, often in the axils of fallen leaves ; rhachis tomentose. Flowers reddish purple or salmon-red, 4-merous. Lower receptacle 5 mm. long, densely pilose ; upper receptacle campanulate or infundibuliform, slightly constricted above the part containing the disk, 12 mm. long, 3–6 mm. in diameter, sericeous-pubescent. Sepals broadly triangular, 2–2·5 mm. long. Petals oblong-spathulate or oblong-elliptic, 4–5·5 mm. long, 1–1·7 mm. wide, pubescent outside, glabrous inside. Stamen-filaments 18–22 mm. long, 2-seriate ; anthers red, 1·5 mm. long. Disk 2–3 mm. in diameter, glabrous except for a pilose margin, which is not free. Style ± 26 mm. long. Fruit 4-winged, elliptic in outline, 3·5–5 cm. long, 2·5–3 cm. wide, apex and base truncate velutinous ; apical peg up to 1 mm. long ; wings up to 10 mm. wide ; stipe up to 10 mm. long.

Tanganyika.   Tunduru District: Mitawatawa, Feb. 1957, *B. D. Nicholson* 119 ! ; Masasi, *Gillman* 1086 ! ; Newala District : Makonde Plateau, *Gillman* 1218 !
Distr.   T8 ; northern Mozambique
Hab.   Deciduous thicket ; ± 500 m.

Syn.   [*C. lukafuense* sensu Liben in F.C.B., Combr.: 27 (1968), pro specim. Mozambiquense, *non* De Wild.]

Note.   This species is very close to *C. lukafuense* De Wild. which differs in having subulate sepals 5–8 mm. long and slightly larger petals up to 7 mm. long ; there is also the considerable discontinuity in distribution which also suggests that the two species should be kept separate, at least for the present, until more material of this very rare species becomes available.

38. C. aculeatum *Vent.*, Choix. Pl. : sub. t. 58 (1808) ; Laws. in F.T.A. 2 : 423 (1871) ; Engl. & Diels in E.M. 3 : 93, t. 27/A (1899) ; Burtt Davy, Checklists Brit. Emp. 1, Uganda : 35 (1935) ; T.T.C.L. : 133 (1949) ; K.T.S. : 141 (1961) ; Liben in F.C.B., Combr.: 25, fig. 2/B (1968). Type : Senegal, *Roussillon* (P-JU, holo., IDC microfiche)

A scandent shrub or in the absence of support a compact or rambling shrub up to 4 m. high ; young branches grey to rufous pubescent, older shoots with yellow-brown bark. Leaves alternate or subopposite ; lamina broadly elliptic to obovate, up to 7 cm. long and 5 cm. wide, apex varying from retuse to shortly acuminate, base cuneate, lightly to densely pubescent on both surfaces ; lateral nerves 4–6 pairs, rather prominent beneath ; petiole 0·1–1 cm. or more long, usually persistent and forming recurved spines up to 1·7 cm. long. Inflorescence of short axillary racemes. Flowers yellowish white, fragrant. Lower receptacle 6–8 mm. long, constricted above and below the ovary, tomentose ; upper receptacle urceolate-campanulate, (3–)4·5–6 mm. long, 3–4 mm. in diameter, pubescent. Sepals deltate, sometimes attenuated. Petals oblanceolate, 4–6 mm. long, 1–2 mm. wide, lightly pubescent externally. Stamen-filaments 4–9 mm. long, 2-seriate ; anthers 0·7 mm. long. Disk without a free margin, pubescent. Fruit (fig. 7/38, p. 51), 5-winged, obovate in outline, 1·5–2·2 cm. long, 1·5–2·3 cm. wide, apex emarginate; body shortly pubescent; wings papyraceous, yellow-brown; apical peg short; stipe 6–12 mm. long.

UGANDA. W. Nile District: Koboko, Mar. 1934, *Tothill* 2528!; Karamoja District: Amudat, Feb. 1958, *Tweedie* 1489!; Mbale District: Nabiswa, 17 Jan. 1955, *Norman* 246!

KENYA. N. Frontier Province: Archers Post, Nov. 1965, *Makin* 264!; Turkana District: Oropoi, Feb. 1965, *Newbould* 7014!; Machakos District: Athi R., Kayata [Kagata] Estate, *Gardner* in *E.A.H.* 95/60/2!

TANGANYIKA. Moshi District: Himo, 28 Apr. 1968, *Bigger* 1837!; Tanga District: Nyika steppe, 7 Mar. 1893, *Holst* 2408!; Lushoto/Tanga District: Umba, Dec. 1892, *S. Smith*!

DISTR. U1–3; K1–4, 6, 7; T2, 3; Senegal east to Somali Republic

HAB. *Acacia, Commiphora* bushland, thicket and wooded grassland, sometimes riverine; 10–1700 m.

SYN. *C. ovale* G. Don in T.L.S. 15: 434 (1827). Type: Ethiopia, without precise locality, *Salt* (BM, holo.)
　　Poivrea aculeata (Vent.) DC., Prodr. 3: 18 (1828)
　　P. ovalis (G. Don) Walp., Rep. Bot. 2: 64 (1843)
　　P. hartmanniana Schweinf., Pl. Nilot. 8, t. 2 (1862), *non Combretum hartmannianum* Schweinf. Type: Sudan, Sennar, *Hartmann* (B, holo. †)
　　Combretum leuconili Schweinf. in Hoehnel, Reise zum Rudolf-See, Append. 2, No. 138 (1892), English translation 2: 364 (1894); P.O.A. C: 292 (1895); Engl. & Diels in E.M. 3: 94 (1899), in synonymy as *C. leuconiloticum* Schweinf. Type: as for *Poivrea hartmanniana* Schweinf.
　　Commiphora holstii Engl., P.O.A. C: 229 (1895), *non Combretum holstii* Engl. (1895). Type: Tanganyika, Nyika steppe, *Holst* 2408 (B, holo. †, K, iso.!)
　　Combretum denhardtiorum Engl. & Diels in E.M. 3: 93, t. 27/B (1899); K.T.S.: 143 (1961). Type: Kenya, Tana R., Massa, *F. Thomas* 114 (B, holo. †, K, iso.!)
　　C. sp. near *C. cinereopetalum* sensu T.S.K., ed. 2: 34 (1936)

39. **C. longispicatum** (*Engl.*) *Engl. & Diels* in E.M. 3: 95 (1899); T.T.C.L.: 133 (1949). Types: Tanganyika, Tabora, *Stuhlmann* 605 (B, syn. †) & Dodoma District, Lake Chaya [Tschaia], *Stuhlmann* 434 (B, syn. †) & Saranda, *Fischer* 254 (B, syn. †, K, fragment!)

A twining and climbing shrub up to 8 m.; rootstock tuberous; branches velutinous when young, becoming glabrescent. Leaves opposite or sub-opposite; lamina ovate to broadly ovate, up to 11 cm. long and 7 cm. wide, apex acutely acuminate, base cordate to subcordate, densely villous when young, less so when mature; lateral nerves 4–8 pairs; petiole 3 mm. long, sometimes developing into a spine 2·5 cm. long bearing a rudimentary leaf. Inflorescences (and infructescences) appearing before the leaves, 1–2 dense secund spikes up to 22 cm. long in the axils of the old leaves. Flowers pink. Lower receptacle 3 mm. long, stout and densely silvery to rufous sericeous pubescent; upper receptacle campanulate, 9–11 mm. long, ± 4 mm. in diameter, densely silky pubescent. Sepals triangular-subulate, ± 2 mm. long. Petals oblong-obovate, 5–6 mm. long, 1·2–1·5 mm. wide, pubescent externally. Stamen-filaments 18–20 mm. long, bright green, biseriate; anthers 1·5–2 mm. long, brownish crimson. Disk-margin not free, pilose. Fruit (fig. 7/39, p. 51, and 8/4, p. 62) 5-winged, ovate to broadly obovate in outline, 2·3–2·8 cm. long, 2–2·2 cm. wide, pubescent; apical peg absent; stipe 4–5 mm. long. Fig. 8, p. 62.

TANGANYIKA. Shinyanga, Aug. 1935, *B. D. Burtt* 5214!; Tabora District: 56 km. from Tabora to Nzega, 25 July 1950, *Bullock* 3014!; Iringa District: Ruaha National Park, Msembe–Mbagi, 4 Aug. 1969, *Greenway & Kanuri* 13686! & 19 Mar. 1970, *Greenway & Kanuri* 14151!

DISTR. T1, 4–7; not known elsewhere

HAB. Deciduous bushland and woodland; 900–1600 m.

SYN. *Cacoucia longispicata* Engl. in P.O.A. C: 293 (1895)
　　Combretum houyanum Mildbr. in E.J. 51: 231 (1914); T.T.C.L.: 133 (1949). Type: Tanganyika, near Kilosa, *Houy* in *Meyer* 1147 (B, holo. †)

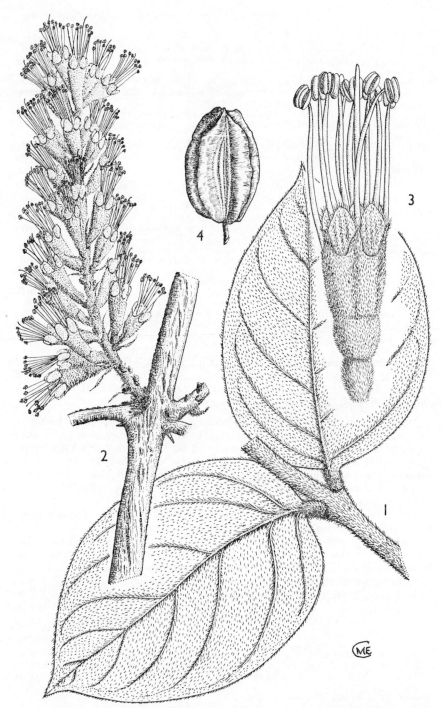

FIG. 8. *COMBRETUM LONGISPICATUM*—**1,** leafy branchlet, × 1; **2,** flowering branchlet, × 1; **3,** flower, × 3; **4,** fruit, × 1. 1, from *B. D. Burtt* 1030; 2, 3, from *Bullock* 3014; 4, from *Greenway* 3668. Drawn by Mrs. M. E. Church.

NOTE. More fruiting and vegetative material of this rather interesting species is required.

40. **C. mossambicense** (*Klotzsch*) *Engl.*, P.O.A. C: 292 (1895); Engl. & Diels in E.M. 3: 98, t. 26/B (1899); T.T.C.L.: 134 (1949); Liben in F.C.B., Combr.: 26 (1968); Exell in Kirkia 7: 219 (1970). Type: Mozambique, Manica e Sofala, Sena, *Peters* (B, holo. †, BM, fragment !)

Small tree, shrub or woody climber; branchlets usually pale pinkish brown, pubescent at first, often soon (but not always) glabrescent. Leaves opposite or subopposite; lamina elliptic to elliptic-oblong or rarely subcircular, up to 20 cm. long and 11 cm. wide (often much smaller, ± 10 cm. long, 5 cm. wide), apex usually acuminate, base rounded to cordate, hairy when young, usually glabrescent but sometimes retaining a dense indumentum; lateral nerves 5–9 pairs; petiole up to 7 mm. long, the base forming a blunt spinal process. Inflorescence of axillary spikes, sometimes subcapitate, up to 8 cm. long, often in the axils of fallen leaves; rhachis tomentose. Flowers (fig. 6/40, p. 50) (4–)5-merous, white or pinkish, usually appearing before the leaves. Lower receptacle ± 5 mm. long, tomentose with short often spreading hairs, often somewhat constricted above and below the ovary; upper receptacle with upper part broadly infundibuliform, lower part surrounding the disk subglobose, slightly constricted between the 2 parts, up to 9(–11) mm. long, 4 mm. in diameter, hairy. Sepals triangular, 2 mm. long. Petals elliptic, 7–9 mm. long, 2–3·5 mm. wide, clawed, pilose externally. Stamen-filaments 16–17 mm. long; anthers 1·7–1·8 mm. long, orange-red. Disk glabrous; margin not free. Fruit (fig. 7/40, p. 51) (4–)5-winged, elliptic to subcircular in outline, 2–3(–4·5) cm. long, 2–2·5(–3) cm. wide, pubescent; apical peg very short; wings up to 10 mm. wide; stipe 4–6(–10) mm. long.

UGANDA. Acholi/Bunyoro District: Murchison Falls, *Jex-Blake* in *E.A.H.* 10073 !
KENYA. Machakos District: Chyulu foothills, Noka Mt., 12 June 1938, *V. G. van Someren* 1053 in *C.M.* 7800 !; Kitui District: Mumoni, Aug. 1937, *Gardner* 3688 !; Machakos/Teita District: Tsavo National Park East, 30 Sept. 1965, *Hucks* 439 !
TANGANYIKA. Shinyanga Hill, 3 May 1945, *Greenway* 7400 !; Tabora District: Talikwa, 8 July 1954, *Joseph* 4013 !; Dodoma District: Kazikazi, 31 July 1931, *B. D. Burtt* 3339 !
DISTR. U1/2; K4, 7; T1, 2, 4–8; Zaire, Malawi, Mozambique, Zambia, Rhodesia, Botswana, South West Africa and South Africa (Transvaal)
HAB. Riverine forest, *Brachystegia* woodland, deciduous and secondary bushland, wooded grassland; 700–1600 m.

SYN. *Poivrea mossambicensis* Klotzsch in Peters, Reise Mossamb., Bot. 1: 78, t. 13 (1861)
[*Combretum constrictum* sensu Laws. in F.T.A. 2: 423 (1871), pro parte quoad syn. *Poivrea mossambicensis, non* (Benth.) Laws.]
C. ukambense Engl., P.O.A. C: 291 (1895); K.T.S.: 147 (1961). Types: Kenya, Kitui, *Hildebrandt* 2615 (B, syn. †), 2824 (B, syn. †, BM, isosyn. !) & 2779 (B, syn. †, BM, K, isosyn. !)
C. trichopetalum Engl., P.O.A. C: 292 (1895). Types: Tanganyika, Mwanza District, Makolo, *Stuhlmann* 722 & Bukumbi, *Stuhlmann* 822 & without precise locality, *Fischer* 250 (all B, syn. †)
C. migeodii Exell in J.B. 67: 48 (1929); T.T.C.L.: 137 (1949). Type: Tanganyika, Lindi District, Tendaguru, *Migeod* 269 (BM, holo. !)

NOTE. *C. holstii* and *C. goetzei* are closely related to *C. mossambicense*, differing principally in the indumentum of the flowers, and ultimately may be regarded as no more than subspecies or varieties of this species (see Exell in Kirkia 7: 222 (1970)). Occasional specimens seem to be somewhat intermediate, e.g. *R. M. Davies* 1002 ! from Lushoto District, Mashewa, is referred here to *C. holstii*, but has some hairs on the upper receptacle, approaching the situation in *C. mossambicense*.

Exceptionally large fruits, 4·5 cm. long, 3 cm. wide, in *Michelmore* 1089 from Milepa near Lake Rukwa are reminiscent of *C. lasiocarpum* Engl. & Diels, which is known only from Mozambique and differs from *C. mossambicense* in having longer fruits and longer spikes, otherwise being very similar. In view of the distance between Milepa and the nearest known locality of *C. lasiocarpum*, I hesitate in referring the specimen to that species.

41. **C. goetzei** *Engl. & Diels* in E.M. 3: 97 (1899); T.T.C.L.: 134 (1949); Exell in Kirkia 7: 221 (1970). Type: Tanganyika, Morogoro District, Kisaki, Wegu Mt., *Goetze* 360 (B, holo. †)

Shrub, sometimes scandent, up to 10 m. high; branchlets greyish or reddish, pubescent when young but soon glabrescent. Leaves opposite or subopposite, sometimes on short shoots in the axils of old leaves; lamina subcoriaceous, oblong-elliptic to obovate, up to 8·5 cm. long and 4 cm. wide, apex rounded or acuminate, base rounded to subcordate, almost glabrous when mature; lateral nerves 6–7 pairs; petiole up to 5 mm. long, often forming a blunt spinal process. Inflorescence of terminal and lateral spikes up to 10 cm. long; rhachis with dense macroscopic glandular hairs. Flowers white, pink or yellowish red, generally appearing before the new leaves. Lower receptacle glandular, ± 5 mm. long; upper receptacle with broadly infundibuliform upper part, lower part surrounding the disk subglobose, slightly constricted between the 2 parts, (4–)6–10 mm. long, 3–4·5 mm. in diameter, with dense macroscopic glandular hairs present but otherwise nearly glabrous. Sepals triangular, 2·5 mm. long. Petals 7–9 mm. long, 2–2·5 mm. wide. Stamen-filaments 18–19 mm. long; anthers 1·8 mm. long. Disk glabrous except for pilose margin, which is not free. Fruit (fig. 7/41, p. 51) 5-winged, elliptic or ovate-elliptic or broadly elliptic in outline, 3 cm. long, 2·5 cm. wide, glabrous, somewhat glutinous; apical peg ± 0·5 mm. long; wings up to 10 mm. wide; stipe 2–3 mm. long.

Tanganyika. Kilosa District: Ilonga, *Robson* BH 11!; Morogoro District: Kisaki, 6 Dec. 1933, *B. D. Burtt* 5003! & Kihonda Sisal Estate, Nov. 1949, *Semsei* in *F.H.* 2865!
Distr. **T6**; Mozambique
Hab. Very common in grassland (*Semsei* in *F.H.* 2865), no further information available; 500–600 m.

Note. See note under 40, *C. mossambicense*.

42. **C. holstii** *Engl.*, P.O.A. C: 291 (1895); Engl. & Diels in E.M. 3: 95, t. 28/A (1899); T.S.K., ed. 2: 33 (1936); T.T.C.L.: 134 (1949); Liben in F.C.B., Combr.: 24 (1968); Exell in Kirkia 7: 222 (1970). Types: Tanganyika, Lushoto District, Korogwe, *Holst* 3978 (B, syn. †) & Lutindi, *Holst* 3487 (B, syn. †, K, isosyn. !)

Shrub and woody climber to 7 m.; branchlets pale, pinkish brown, glabrous. Leaves opposite or subopposite; lamina chartaceous, oblong-elliptic, up to 15(–20) cm. long and 7·5(–8·5) cm. wide, apex acuminate, base rounded to cordate, minutely glandular-pustulate but otherwise glabrous; lateral nerves 7–9(–11) pairs; petiole up to 10 mm. long, with glabrous basal part forming a curved spine. Inflorescences of axillary spikes up to 14 cm. long, often in the axils of fallen leaves; rhachis glabrous. Flowers crimson or purplish red, often appearing with the leaves. Lower receptacle ± 5 mm. long, glabrous except sometimes for glandular hairs; upper receptacle with broadly infundibuliform upper part and subglobose lower part surrounding the disk, slightly constricted between the 2 parts, 10–15 mm. long, 5–6 mm. in diameter, glabrous or sparsely glandular-hairy. Sepals deltate, 2–2·5 mm. long. Petals narrowly ovate-elliptic, up to 6 mm. long, 2–2·8 mm. wide, clawed, pubescent externally. Stamen-filaments 17–18 mm. long; anthers 2 mm. long. Disk glabrous except for pubescent margin, which is not free. Fruit (fig. 7/42, p. 51) 5-winged, broadly elliptic to subcircular in outline, up to 3·5 cm. long, 3 cm. wide, glabrous, shiny; apical peg very short; wings up to 10 mm. wide; stipe up to 9 mm. long.

TANGANYIKA. Lushoto District: Mombo Forest Reserve, Aug. 1955, *Mgaza* 61!; Handeni District: Kideleko, 27 July 1965, *Archbold* 554!; Ulanga District: Ifakara, 24 June 1959, *Haerdi* 248/87!
ZANZIBAR. Zanzibar I., Makunduchi road, 19 Aug. & 6 Oct. 1951, *R. O. Williams* 79 & 94!
DISTR. **T3**, 6–8; **Z**; Zaire, Mozambique and Angola
HAB. A wide range of habitats from open grassland to riverine forest and forest margins; sea-level to 900 m.

SYN. [*C. ukambense* sensu T.T.C.L.: 135 (1949), *non* Engl. (1895)]

NOTE. See note under 40, *C. mossambicense*.

43. **C. constrictum** (*Benth.*) *Laws.* in F.T.A. 2: 423 (1871), excl. syn. *Poivrea mossambicensis* Klotzsch et formae; Engl. & Diels in E.M. 3: 99, t. 26/A (1899); T.T.C.L.: 133 (1949); K.T.S.: 143 (1961); Exell in Kirkia 7: 223 (1970). Type: Nigeria, R. Niger [Quorra] at Idah [Aboh], *Vogel* 39 (K, holo.!)

Shrub or climber; branchlets glabrescent. Leaves subopposite; lamina chartaceous to subcoriaceous, oblong to oblong-elliptic, up to 12 cm. long and 6·5 cm. wide, apex rounded or acuminate, base rounded, sometimes tomentose, more usually glabrous or nearly so; lateral nerves 6–9 pairs; petiole 4–6 mm. long, eventually forming a blunt and slightly curved spine at the base. Inflorescence of subcapitate terminal or axillary spikes up to 6 cm. long; rhachis glabrous or tomentose. Flower buds unusually large, clavate. Flowers (fig. 6/43, p. 50) red or pinkish. Lower receptacle glabrous or tomentose; upper receptacle tubular, broadly infundibuliform at the apex, 10–18 mm. long, 7 mm. in diameter at the apex, constricted in the middle for 4–7 mm. in length, 1·5 mm. in diameter, glabrous or tomentose. Sepals triangular, 3 mm. long. Petals narrowly elliptic, 7–8 mm. long, 2·5 mm. wide, clawed, pilose. Stamen-filaments up to 25 mm. long, 2-seriately inserted on the apical part of the upper receptacle; anthers 0·8 mm. long. Disk 2 mm. in diameter; margin not free, glabrous. Fruit (fig. 7/43, p. 51) 5-angled, ellipsoid, sessile, ± 2·5 cm. long, 1·2–1·5 cm. in diameter, glabrous; apical peg absent.

KENYA. Kwale District: Kirao, Jan. 1930, *R. M. Graham* in *F. D.* 2243!; Tana R. District: Garissa, 26 Sept. 1957, *Greenway* 9233! & Galole, 23 Mar. 1965, *Makin* 15!
TANGANYIKA. Pangani District: Bushiri Estate, 1 May 1950, *Faulkner* 578!; Mpanda District: Kabungu, 20 July 1948, *Semsei* 163!; Morogoro District: Turiani, 23 Nov. 1955, *Milne-Redhead & Taylor* 7362!
ZANZIBAR. Zanzibar I., without precise locality, *Speke* 11! & Mazizini [Massazine], 20 May 1959, *Faulkner* 2263!
DISTR. **K**1, 4, 7; **T**3, 4, 6, 8; **Z**; Nigeria, ? Somali Republic, Mozambique
HAB. Tidal and seasonal swamps and riverine forest; sea-level to 1200 m.

SYN. *Poivrea constricta* Benth. in Hook., Niger Fl.: 337 (1849)
     Combretum infundibuliforme Engl., P.O.A. C: 292 (1895); T.S.K., ed. 2: 33 (1936). Types: Zanzibar I., *Stuhlmann* 633 & 945 & Tanganyika, Pangani, *Stuhlmann* 509 & 944 (all B, syn. †) & Bagamoyo District, Wami and Ruvu [Kingani] rivers, *Hildebrandt* 1176 (B, syn. †, K, isosyn.!)
     C. constrictum (Benth.) Laws. var. *tomentellum* Engl. & Diels in E.M. 3: 100 (1899); T.T.C.L.: 134 (1949). Type: Tanganyika, Rufiji District, Mtanza [Mtansa], *Goetze* 51 (B, holo. †, K, iso.!)
     ? *C. bussei* Engl. & Diels in E.J. 39: 509 (1907); T.T.C.L.: 133 (1949). Type: Tanganyika, Lindi, *Busse* 2429 (B, holo. †, EA, iso.)
     ? *C. constrictum* (Benth.) Laws. var. *somalense* Pampan. in Bull. Soc. Bot. Ital. 1915: 12 (1915). Type: Somali Republic (S.), R. Juba, Giumbo, *Pampanini* 236 & Zingibar, *Pampanini* 370 & R. Shebelli, Burbisciaaro-Gascianle, *Pampanini* 1310 (all FI, syn.)
     C. sp. sensu U.O.P.Z.: 207 (1949)

Imperfectly known species

44. C. sp. A

Small tree or subscandent shrub. Leaves opposite; lamina coriaceous, elliptic, 7–11·5 cm. long, 3–5·2 cm. wide, apex apiculate, base cuneate, often minutely pubescent on the midrib, sometimes with domatia in the nerve-axils, otherwise glabrous except for the dense but not contiguous scales; lateral nerves (and midrib) impressed above, prominent beneath, 6–8 pairs; petiole 3–5 mm. long. Inflorescence axillary. Flowers not seen. Fruit subcircular in outline, apex retuse, base cordate, 1·6–1·8 cm. long and wide; body and wings rufous lepidote; apical peg absent; stipe 2–3 mm. long. Scales 100 μ in diameter, cells numerous.

Tanganyika. Uzaramo District: Pugu Forest Reserve, June 1954, *Semsei* 1749! & June 1954, *K. A. Mgaza* 2! & May 1965, *Procter* 2986! & Soga, 30 July 1939, *Vaughan* 2864!
Distr. T6; not known elsewhere
Hab. Lowland dry evergreen forest; ± 100 m.

Note. Flowering material of this very local species is required before it can be correctly placed in a section, but on scale characters it seems most likely to belong to either section *Hypocrateropsis* or section *Campestres* Engl. & Diels.

45. C. sp. B

Evergreen tree 17 m. high. Leaves opposite; lamina coriaceous, ovate to oblong-ovate or elliptic, 12–14 cm. long, 6·5–7 cm. wide, apex obtuse to apiculate, base cuneate, glabrous and somewhat glutinous, no scales seen; lateral nerves subprominent above, prominent beneath, 4–6 pairs; petiole 0·8–1·4 cm. long. Inflorescences solitary axillary spikes 5–6·5 cm. long. Flowers white, 4-merous. Lower receptacle 3 mm. long, densely shortly pubescent; upper receptacle cup-shaped, 3 mm. long, 2·5 mm. wide, shortly pubescent. Sepals deltate, 0·7 mm. long. Petals broadly obovate, 1·2 mm. long and wide. Stamens immature. Disk pilose. Fruit not seen.

Tanganyika. Lushoto/Tanga District: Mangubu–Kiwanda, Sigi R., 21 Oct. 1936, *Greenway* 4686!
Distr. T3; known only from the one gathering
Hab. Riverine forest; 300 m.

Note. Section not known. More material with mature flowers and fruit is required.

46. C. lindense *Exell & Mildbr.* in N.B.G.B. 14: 105 (1938); T.T.C.L.: 134 (1949). Type: Tanganyika, Lindi District, Makonde Plateau, *Schlieben* 6473 (B, holo. †)

Climbing shrub. Leaf-lamina elliptic, 4–6 cm. long, 1·5–3 cm. wide, glabrous; lateral nerves 5 pairs, prominent, reticulation prominent, domatia present in the nerve-axils; petiole 5–8 mm. long. Inflorescence of few spikes 6–8 cm. long, forming axillary panicles. Lower receptacle 1·5 mm. long; upper receptacle broadly cup-shaped, glabrous. Sepals semielliptic-deltoid, ± 1 mm. long, acute. Petals white, spathulate, 2 mm. long, 0·8 mm. wide, clawed, glabrous. Stamen-filaments ± 2 mm. long. Disk with pilose margin.

Tanganyika. Lindi District: Makonde Plateau, 3 May 1935, *Schlieben* 6473
Distr. T8
Hab. Bushland; 500 m.

Note. From the description it is not possible to establish the affinity of this species. Exell & Mildbraed suggest an affinity with sections *Hypocrateropsis* and *Paucinerves* Engl. & Diels but without the conspicuous scales of these groups.

2. QUISQUALIS*

L., Sp. Pl., ed. 2 : 556 (1762); Engl. & Diels in E.M. 4 : 5 (1900); Exell in
J.B. 69 : 117 (1931), excl. *Q. hensii*; Exell & Stace in Bol. Soc. Brot., sér. 2,
38 : 139 (1964) & 40 : 18 (1966)

Udani Adans., Fam. Pl. 2 : (22) (1763), *nom. illegit.*
Kleinia Crantz, Inst. 2 : 488 (1766), *non* Jacq. (1763)
Spalanthus Jack, Mal. Misc. 2(7) : 55 (1822), as " *Sphalanthus* "
Campylogyne Hemsl. in Hook., Ic. Pl., t. 2550 (1897)

Woody climbers; stalked glands present, peltate scales absent. Leaves
opposite or subopposite, entire, glabrous or hairy; bases of the petioles
persisting as spines after leaf-fall. Flowers ♂̄, regular or slightly zygomorphic,
5-merous, in elongated, usually unbranched, terminal or axillary bracteate
spikes. Receptacle hairy or nearly glabrous; upper receptacle infundibuliform, elongate-infundibuliform or trumpet-shaped. Sepals triangular,
sometimes with filiform tips. Petals 2–20 mm. long, exceeding the sepals and
enlarging during anthesis. Stamens 10, biseriate, inserted near the mouth of
the upper receptacle and not exserted beyond the petals. Disk narrowly
tubular or absent. Style adnate for part of its length to the inner wall of the
upper receptacle. Fruit 5-winged, dehiscent or indehiscent.

Probably about 16 species, 8 in tropical Asia and 8 in tropical and subtropical Africa.

The genus is separated from *Combretum* on account of the style being adnate for about
half its length to the upper receptacle, together with the non-exsertion of the stamens.
Quisqualis appears to be related to *Combretum* subgen. *Cacoucia* (Exell & Stace in Bol.
Soc. Brot., sér. 2, 40 : 12 (1966)).

The Congo species *Q. falcata* Hiern var. *mussaendiflora* (Engl. & Diels) Liben with
brilliant red bracts is cultivated in Nairobi.

Upper receptacle 6–8 cm. long; fruit ovate-
    elliptic in outline with thick narrow wings   .   **1.** *Q. indica*
Upper receptacle up to 2·5 cm. long; fruit sub-
    circular in outline with thin wide wings   .   **2.** *Q. littorea*

1. **Q. indica** *L.*, Sp. Pl., ed. 2 : 556 (1762); Laws. in F.T.A. 2 : 435 (1871);
Engl. & Diels in E.M. 4 : 5, fig. 3 (1900); T.T.C.L. : 142 (1949); U.O.P.Z : 429,
fig. (1949); Liben in F.C.B., Combr. : 88 (1968). Type: specimen received
from Clifford of unknown origin, *Linnean Herb.* 553.1 (LINN, lecto. *fide*
Exell in J.B. 69 : 124 (1931))

Woody climber; young branchlets tomentose to sparsely pubescent, rarely
sparsely glandular. Leaves opposite or subopposite; lamina papyraceous,
elliptic or oblong-elliptic, 8–14·5 cm. long, 3·5–9 cm. wide, apex acuminate or
subcaudate, base rounded or subcordate, tomentose to nearly glabrous,
minutely verruculose; lateral nerves 5–7 pairs, domatia sometimes present;
petiole up to 10 mm. long, the base sometimes persisting and forming a spine.
Inflorescence terminal and axillary spikes 2–5(–10) cm. long, sometimes
forming a leafy panicle; bracts lanceolate-acuminate or elliptic, 6–10 mm.
long, 1–3 mm. wide. Flowers fragrant. Lower receptacle 3–4 mm. long,
pubescent to sericeous-tomentose; upper receptacle narrowly tubular,
expanding slightly at the apex, 6–8 cm. long, tomentose to pubescent.
Sepals triangular, 1–3 mm. long, acute. Petals imbricate in bud, white
becoming dark red on the inner face, enlarging at anthesis, oblong to oblong-
obovate, up to 25 mm. long, 13 mm. wide, apex acute (in the Flora area) or

* *Quisqualis* is the latinized form of the Dutch name " hoedanig " which translated
means " how what ", and is a pun on the Malayan vernacular name " udani " (Stearn,
Gard. Dict. Plant Names: 269 (1972)).

obtuse, shortly clawed. Stamen-filaments 7–8 mm. long; anthers 0·9 mm. long. Style with upper part free for 10–20 mm. Fruit (from West African material) ovate-elliptic in outline, 2·5–4 cm. long, 0·75–1·25 cm. wide, appressed pubescent to glabrous; wings 1–2 mm. wide, stout; stipe 0·5–1 mm. long.

TANGANYIKA. Mpwapwa, 25 Apr. 1932, *B. D. Burtt* 3875 !; Rufiji, 10 Feb. 1931, *Musk* 32 !; Iringa District: Mgole, 16 Jan. 1952, *Wigg* 992 !
DISTR. T5–7; reputed to be a native of Asia, but probably also native in tropical Africa, now widely cultivated in the tropics, often naturalized
HAB. *Acacia* bushland, *Brachystegia* woodland, hillslopes and along stream banks; 900–1350 m.

NOTE. This is a very variable species. All the indigenous specimens from the Flora area are unusual in the petals being rather abruptly acute. In Asia the petals are usually obtuse, although specimens do occur where the apex is acute. In Africa a mixture of both forms occurs. There are other minor differences between Asiatic and African specimens, such as the leaf being more strongly verruculose in Africa; in Asia the stamens are shortly exserted at maturity but apparently not so in Africa. However, before attempting any taxonomic distinction it is highly desirable that the maturation of the flowers should be carefully observed for plants from both Africa and Asia since it is suspected that the stamen-filaments may, like the petals, enlarge at anthesis. The variation in fruit structure should also be studied.

    Q. indica has previously been regarded as a native of Asia, and all the African specimens have been regarded as naturalized escapes. This is now regarded as questionable since *Goetze* 454 (K !) from Iringa District, collected in 1898, is from an area that was quite unsettled at that time. Characteristic Asiatic forms have been cultivated, e.g. Nairobi Arboretum, *G. R. Williams* 396 ! & 529 ! and Tanga, *Greenway* 1821 !

2. **Q. littorea** (*Engl.*) *Exell* in J.B. 69: 121 (1931); T.S.K., ed. 2: 34 (1936); T.T.C.L.: 142 (1949); K.T.S.: 149 (1961). Types: Tanganyika, Tanga, *Holst* 4034 (B, syn. †) & *Holst* 2061 (B, syn. †, K, isosyn. !) & *Volkens* 147 (B, syn. †, BM, K, isosyn. !)*

Scrambling shrub; branchlets lightly golden pubescent, becoming glabrescent, older branches reddish brown; bark peeling in strips. Leaves opposite or subopposite; lamina ovate, up to 8 cm. long and 3·2 cm. wide, apex acute, base cuneate, sparsely pubescent on the venation; midrib prominent beneath; lateral nerves 3–4 pairs; petiole up to 8 mm. long, the base sometimes persisting as a short and blunt spine. Inflorescence of terminal and axillary bracteate spikes up to 14 cm. long (shorter for the axillary spikes); rhachis pubescent; bracts narrowly ovate, 15 mm. long, 3 mm. wide, apex acute, base attenuated. Lower receptacle 3–3·5 mm. long, densely pubescent; upper receptacle narrowly tubular but widening towards the apex, up to 2·5 cm. long, lightly pubescent. Sepals narrowly triangular, up to 1 mm. long, apex attenuated. Petals oblong-ovate, 6 mm. long, 3 mm. wide, apex rounded, base cuneate. Stamen-filaments 4–6 mm. long; anthers 0·7 mm. long. Style 24 mm. long, adnate to the wall of the upper receptacle for 14 mm. long. Fruit subcircular in outline, up to 3·2 cm. long and 2·5 cm. wide, very lightly and shortly pubescent; wings 1·2 cm. wide, papery; stipe up to 6 mm. long, stout. Fig. 9.

KENYA. Kwale District: Mwele Mdogo Forest, 6 Feb. 1953, *Drummond & Hemsley* 1151 !; Kilifi District: Sokoke Forest, 19 Mar. 1954, *Trump* 76 ! & Sabaki, 6 Nov. 1961, *Polhill & Paulo* 721 !

* No collector is mentioned by Engler in his original description, but in E.M. 3: 87 (1899) four collections are cited. Of these *Holst* 4034 and 2061 and *Volkens* 147, all from the Tanga area and collected in 1892–3, were almost certainly seen by him for his original description, whereas *Heinsen* 126, collected in 1895, would have been too late.

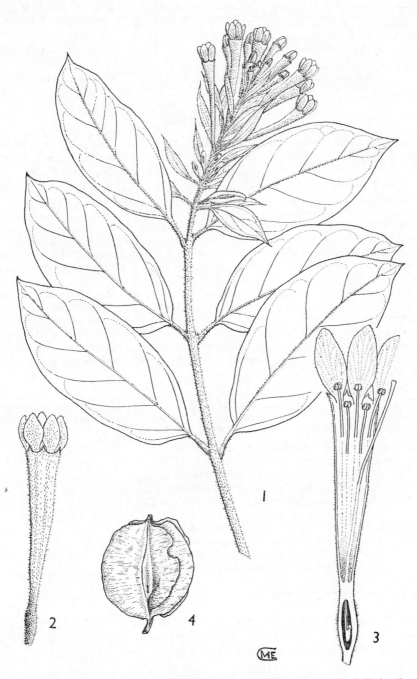

FIG. 9. *QUISQUALIS LITTOREA*—**1,** flowering branchlet, × 1; **2,** flower, × 2; **3,** longitudinal section of flower, × 3; **4,** fruit, × 1. 1–3, from *Drummond & Hemsley* 1151; 4, from *Bally* 5769. Drawn by Mrs. M. E. Church.

TANGANYIKA. Tanga District: Kange Estate, 16 Jan. 1952, *Faulkner* 842! & Kange,
    6 Dec. 1935, *Greenway* 4230!; Uzaramo District: Dar es Salaam, 11 Nov. 1965,
    McCusker 11!
DISTR. **K7**; **T3**, 6; Somali Republic
HAB. Locally abundant in coastal bushland and at forest margins; sea-level to 70 m.

SYN. *Cacoucia littorea* Engl., P.O.A. C: 293 (1895)
        Combretum littoreum (Engl.) Engl. & Diels in E.M. 3: 87, t. 25/D (1899)

3. PTELEOPSIS

Engl. [in Abh. Preuss. Akad. Wiss. 1894: 25 (1894), *nom. nud.*], P.O.A. C:
293 (1895); Engl. & Diels in E.M. 4: 2 (1900); Exell & Stace in Bol. Soc. Brot.,
    sér. 2, 40: 20 (1966); Exell in Kirkia 7: 225 (1970)

Small or medium-sized trees or occasionally shrubs, without scales or
stalked glands. Leaves opposite or subopposite, petiolate, almost glabrous or
hairy. Flowers andromonoecious ($\female$ and $\male$ in the same inflorescence),
(4)5-merous, pedicellate, in terminal and/or axillary or extra-axillary subcapitate racemes. Upper receptacle campanulate, joined to the lower by a
slender stalk-like region; lower receptacle somewhat flattened. Hermaphrodite flowers: sepals deltate, little developed; petals (4)5, usually somewhat
obovate; stamens (8)10, 2-seriate in E. African species; disk pilose with a
short free margin; style not expanded at the apex. Male flowers: usually
towards the base of the inflorescence, similar to the $\female$ flowers but with the
ovary not developing and with a slender stalk replacing the lower receptacle
(perhaps a true pedicel only towards the base); style present or vestigial.
Fruit 2–5-winged, wings often decurrent into the comparatively long slender
stipe. Cotyledons (where known) unfolding spirally and borne above soil-
level.

Probably about 9 species in tropical Africa.

The genus is intermediate in many characters between *Combretum* and *Terminalia*
and is placed by Exell & Stace (*loc. cit.*) in the subtribe *Pteleopsidinae* of the tribe
Combreteae between the subtribes *Combretinae* and *Terminaliinae*.

Lower receptacle glabrous or nearly so; fruit
    2–4(–5)-winged:
    Fruit 2–3(–4 or very rarely –5)-winged, rarely
        with an apical peg, which if present very
        short; leaves generally shiny above, usually
        acuminate; petiole up to 10 mm. long   .   1. *P. myrtifolia*
    Fruit 3–4(–5)-winged; apical peg usually ± 1
        mm. long; leaves usually (but not always) ±
        matt on the upper surface, often mucronate
        or apiculate but not acuminate; petiole up to
        6 mm. long   .   .   .   .   .   .   2. *P. anisoptera*
Lower receptacle pubescent; fruit 4-winged   .   .   3. *P. tetraptera*

1. **P. myrtifolia** (*Laws.*) *Engl. & Diels* in E.M. 4: 4, t. 1/B (1900); T.T.C.L.:
142 (1949); F.F.N.R.: 287, fig. 50/H (1962); Exell in Kirkia 7: 225 (1970).
Types: Mozambique, Lupata and Tete, *Kirk* (K, syn.!)

Tree up to 30 m. high (*Semsei* 847), more often 8–20 m., sometimes a shrub;
wood red, very hard; bark grey or yellow-grey; young branchlets reddish
brown, often pendulous, pubescent at first, soon glabrescent. Leaves opposite or subopposite; lamina elliptic to very narrowly elliptic or obovate-
elliptic, up to 9·5 cm. long and 3 cm. wide, apex slightly acute or bluntly
acuminate, base cuneate, dark green and usually shiny above, usually

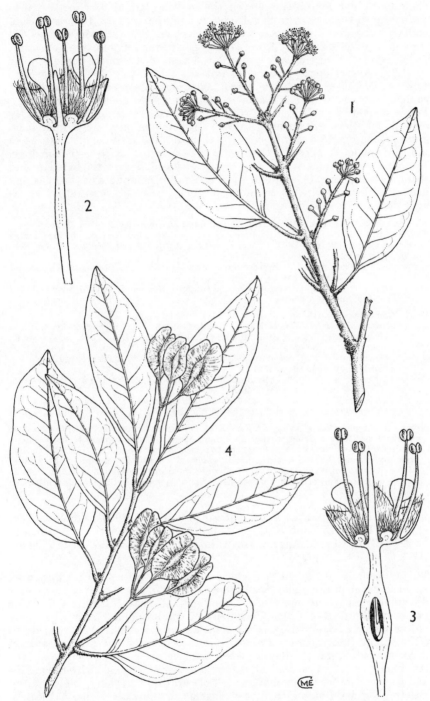

FIG. 10. *PTELEOPSIS MYRTIFOLIA*—**1,** flowering branchlet, × 1; **2,** longitudinal section of male flower, × 6; **3,** longitudinal section of hermaphrodite flower, × 6; **4,** fruiting branchlet, × 1. All from *Fundi* 30. Drawn by Mrs. M. E. Church.

glabrous except for pubescence on the midrib; lateral nerves 6–9 pairs; petiole up to 1 cm. long, rather slender, usually pubescent. Inflorescence of axillary subcapitate racemes up to 4·5 cm. long. Flowers white or yellow, fragrant; ♀ flowers usually towards the apex of the inflorescence, 5-merous, pedicellate; ♂ flowers usually towards the base of the inflorescence, similar to the ♀ flowers (style present) except that the ovary does not develop. Lower receptacle ± 5 mm. long, slender, glabrous; upper receptacle shallowly campanulate, 2·5–3 mm. in diameter, glabrous. Sepals shallowly triangular. Petals obovate to subcircular, 1·5–2·5 mm. long, 1·0–2 mm. wide, shortly clawed, glabrous. Stamens 2-seriate, antipetalous ones 4·5–5 mm. long, antisepalous ones 3·5–4 mm. long; anthers 0·7–0·8 mm. long. Disk 2·5–3 mm. in diameter, pilose, with a short free margin. Style 5–6 mm. long. Fruit 2–3-, sometimes 4- and very rarely 5-winged, very variable in size and shape, 1–3 cm. long, 0·5–1·8 cm. wide, generally emarginate at the apex and usually without or with a very short apical peg, base decurrent into the stipe and often oblique; stipe up to 15 mm. long, very slender. Fig. 10, p. 71.

KENYA. Kwale District: Mrima Hill–Lungalunga, 26 June 1970, *Faden* 70/256!
TANGANYIKA. Masai District: Kijungu–Njoge, 27 July 1965, *Leippert* 6056!; Tanga District: Ngole, 9 June 1937, *Greenway* 4942!; Rufiji District: Mohoro–Utete, Feb. 1952, *Procter* 23!
DISTR. **K**7; **T**2, 3, 6, 8; Malawi, Mozambique, Zambia, Rhodesia, Botswana, Angola and South Africa (Natal, Transvaal)
HAB. Dry evergreen and riverine forest, deciduous woodland, coastal bushland and wooded grassland; 0–1600 m.

SYN. *Combretum myrtifolium* Laws. in F.T.A. 2: 431 (1871)
  Pteleopsis variifolia Engl., P.O.A. C: 293 (1895). Types: Tanganyika, Lushoto District, Gombelo, *Holst* 2179 (B, syn. †) & Hosiga, *Holst* 2512 (B, syn. †, K, isosyn.!) & Tanga District, Amboni, *Holst* 2613 (B, syn. †, K, isosyn.!) & Uzaramo District, *Stuhlmann* 6795 & 7043 (B, syn. †) & Kisarawe [Kisserewe], *Stuhlmann* 6193 (B, syn. †)
  P. stenocarpa Engl. & Diels in E.M. 4: 5 (1900). Type: Angola, Huila, Serra Cheila, *Newton* 159 (COI, holo.)
  P. obovata Hutch. in K.B. 1917: 232 (1917). Types: Mozambique, Niassa, R. Messalo [Msalu], *Allen* 72 & 156 & Madanda Forest, *Dawe* 449 (all K, syn.!)

NOTE. The separation of *P. myrtifolia* and *P. anisoptera* is not easy, even the number of wings per fruit varying within one inflorescence. There appears to be no single character that will infallibly separate the two species and it seems highly probable that they hybridize. In the Flora area, however, *P. myrtifolia* is widely distributed along the east coast, whereas *P. anisoptera* is apparently restricted to Ufipa District at the southern end of Lake Tanganyika. The two species require further intensive study.

2. **P. anisoptera** (*Laws.*) *Engl. & Diels* in E.M. 4: 4 (1900); F.F.N.R.: 287, fig. 50/G (1962); Liben in F.C.B., Combr.: 5, fig. 1/A (1968); Exell in Kirkia 7: 227 (1970). Type: Angola, Huila, Lopolo, *Welwitsch* 4374 (LISU, holo., BM!, COI, K!, P, iso.)

Small tree up to 12(–18) m. high or occasionally a shrub; bark grey with longitudinal fissures; young branchlets at first pubescent, soon glabrescent. Leaves opposite, subopposite or sometimes alternate; lamina elliptic or narrowly elliptic or obovate-elliptic, up to 7 cm. long and 3·5 cm. wide, apex not or scarcely acuminate, usually apiculate or mucronate, base cuneate to rounded, at first densely sericeous-pilose, especially beneath, eventually pilose to pubescent or almost glabrous, matt or somewhat shiny above; lateral nerves 4–6 pairs, tending to form a rather acute angle with the midrib and to run parallel to the margin; petiole up to 6 mm. long; buds conical, nearly glabrous, tending to be conspicuous in the axils of the old leaves. Inflorescences of axillary subcapitate racemes up to 3 cm. long. Flowers white, ♀ and ♂ in the same inflorescence, ♂ flowers similar to the ♀ flowers (style present but often shorter), the ovary not developing and pedicellate for 10 mm. Floral characters similar to *P. myrtifolia*. Fruit 3–4(–5)-winged, obovate to elliptic in outline, 1–1·8 cm. long, 0·6–1·2 cm. wide, glabrous, apex usually emarginate and generally with a distinct apical peg 0·5–3 mm. long, base decurrent into the stipe and usually oblique; stipe up to 2 cm. long, slender.

TANGANYIKA. Ufipa District: Kasanga, 20 June 1957, *Richards* 10178!
DISTR. **T4**; Zaire, Zambia, Mozambique, Rhodesia and Angola
HAB. Not known for Flora area, but occurs in *Brachystegia* woodland and *Combretum*, *Commiphora* thickets in Zambia; 840 m.

SYN.   *Combretum anisopterum* Laws. in F.T.A. 2: 429 (1871)
   Pteleopsis ritschardii De Wild., Contr. Fl. Katanga, Suppl. 2: 82 (1929). Types:
      Zaire, Katanga, Mukulakulu, *Ritschard* 1428 & Dilolo, *Sapin* 155 (both BR,
      syn.)

NOTE. Further gatherings of the species are required from E. Africa.

3. **P. tetraptera** *Wickens* in K.B. 25: 182, fig. 3 (1971). Type: Kenya, Kilifi District, N. Giriama Reserve, *Dale* in *F.D.* 3656 (K, holo.!, EA, iso.)

Tree 15 m. or more high; branchlets greyish white, lightly pubescent. Leaves subopposite, somewhat glutinous when young; lamina elliptic, up to 7 cm. long and 3·5 cm. wide, apex obtuse or acuminate, base rounded, sericeous pilose, becoming glabrescent except for the pubescent midrib; lateral nerves 7–9 pairs; petiole ± 3 mm. long, pubescent. Inflorescences of axillary subcapitate racemes up to 2 cm. long. Flowers white or cream, 4- or sometimes 5-merous within the same inflorescence and irrespective of whether ♀ or ♂; ♂ flowers similar to the ♀, but style shorter and tending to be near the base of the inflorescence. Lower receptacle ± 7 mm. long, shortly pilose; upper receptacle shallowly campanulate, 2·5–3 mm. long, shortly pilose at the base. Sepals deltate, 1·5 mm. long. Petals obcordate to inversely reniform, 1·2 mm. long, 1–1·8 mm. wide, glabrous. Stamens 2-seriate; antipetalous filaments 6 mm. long, antisepalous ones 5 mm. long; anthers 0·7 mm. long. Disk 2·5–3 mm. in diameter, pilose with short free margin. Style ± 4 mm. long. Fruit 4-winged, subcircular in outline, 1·2 cm. long and wide, deeply emarginate, base rounded, glabrous; apical peg absent; stipe 7 mm. long, slender, pubescent.

KENYA. Mombasa District: Mowesa, *R. M. Graham* in *F.D.* 1742!; Kilifi District:
   N. Giriama Reserve, Jan. 1937, *Dale* in *F.D.* 3656! & Bamba–Sokoke road, 11 Dec.
   1962, *Dale* 2020A!
TANGANYIKA. Lushoto District: Kabaku Forest, 31 Dec. 1968, *Faulkner* 4181!
DISTR. **K7**; **T3**; not known elsewhere
HAB. Coastal bushland, wooded grassland and lowland dry evergreen forest; sea-level
   to 300 m.

SYN.   *P. sp.* sensu T.S.K., ed. 2: 34 (1936)
   [*P. myrtifolia* sensu K.T.S.: 149 (1961), *non* (Laws.) Engl. & Diels]

4. TERMINALIA

L., Syst. Nat., ed. 12, 2: 674 [err. 638] (1767) & Mant. Pl.: 21 (1767); Engl. & Diels in E.M. 4: 6 (1900); Griffiths in J.L.S. 55: 818 (1959), *nom. conserv.* For full synonymy see Exell in J.B. 69: 125 (1931), excl. syn. *Terminaliopsis* Danguy.

Trees or rarely shrubs, without scales or microscopic stalked glands. Leaves usually spirally arranged, often crowded at the ends of branches, sometimes on short shoots, rarely opposite, petiolate or subsessile, usually entire but sometimes subcrenate, often with 2 or more glands at or near the base of the lamina or on the petiole (but not in native species). Flowers usually ♀ and ♂ in the same inflorescence (rarely all ♀), usually in axillary spikes with ♂ flowers towards the apex and ♀ ones towards the base, rarely in terminal panicles; ♂ flowers stalked, stalks resembling pedicels but corresponding to the lower receptacle with abortion of the ovary, ♀ flowers sessile. Receptacle divided into a lower part (lower receptacle) and an upper part, often scarcely developed, expanding into a shallow cup terminating in the sepals. Petals absent. Stamens usually 10, exserted. Disk intrastaminal. Ovary completely inferior; style free, not expanded at the apex. Fruit very variable in size and shape but usually 2-winged in E. Africa, usually with at least partially sclerenchymatous endocarp. Cotyledons (where known) spirally convolute.

Probably about 200 species in the tropics and subtropics, with about 30 species in Africa.

The descriptions given, with minor alterations, are based mainly on " A Revision of African species of *Terminalia* " by M. E. Griffiths (in J.L.S. 55: 818–907 (1959)). The flowers of *Terminalia* are remarkably uniform throughout the genus and scarcely ever provide any taxonomically useful characters and great reliance must therefore be placed on leaf, bark and fruit characters.

Several introduced species are grown. *T. catappa* L., native of tropical Asia and widely planted throughout the tropics as a shade tree, is common in the coastal regions, e.g. Kenya, Mombasa, *MacNaughton* 120 !, Tanganyika, Amani, *Greenway* 2792 ! and Lindi, *Semsei* 653 !, Zanzibar and Pemba (U.O.P.Z.: 465 (1949)), also occasional in towns around Lake Victoria (Dale, Introd. Trees Uganda: 68 (1953)). It is easily recognized by the very large obovate shortly petiolate leaves, which are subcordate at the base and turn red before falling, and by the large somewhat compressed-ellipsoid fruits ± 6 cm. long. *T. bellirica* (Gaertn.) Roxb., native of tropical Asia, fruits of which are used for tanning and dyes, has been tried in Uganda at the Entebbe Botanic Gardens (*A. S. Thomas* 3002 !; Dale, Introd. Trees Uganda: 67 (1953)) and in Tanganyika at Amani (T.T.C.L.: 143 (1949), as " *belerica* "). It has large leaves like *T. catappa*, but the petiole is 3–8 cm. long and the lamina is not basally subcordate; the fruit is sub-globose to broadly ellipsoid, 5-ridged, 2–2·8 cm. long. *T. arjuna* (Roxb.) Wight & Arn., native of India, with 5-winged fruits, has been grown at Dar es Salaam (*Holtz* 662; T.T.C.L.: 143 (1949)). *T. mantaly* Perrier, native of Madagascar, is grown as an ornamental tree in Nairobi, e.g. *Perkins* in *E.A.H.* 13449 ! & 13753 ! The obovate leaves are borne on spur shoots at the ends of lateral branches and in the angles of the zigzag long shoots; the fruits are ellipsoid, ± 1·8 cm. long, neither winged nor ridged. *T. chebula* Retz., a native of India, Ceylon and Burma, the fruits of which are used for tanning and dyes, is cultivated in Tanganyika at Amani (T.T.C.L.: 143 (1949)). It has large, oblong, ovate or elliptic leaves with petioles 1–3 cm. long; the fruit is broadly ovoid, 5-ridged, 2–3 cm. long.

NOTE. Terminology for the various types of shoot is illustrated in fig. 11, p. 75.

Fruit ellipsoid-ovoid, 10–12(–18) mm. long, 5 mm.
      across, not winged and only obscurely ridged
      (or very narrowly winged in one Madagascan
      species)   .   .   .   .   1. sect. *Fatrea*, p. 75
Fruit 2-winged (rarely 3–4-winged in *Discocarpae*):
   Leaves in fascicles on short spur shoots:
      Leaves subcircular; fruit subcircular, wings
            developed beyond both the style at the
            apex and beyond the insertion of the stipe
            at the base; sepals reflexed   .   .   .   2. sect. *Discocarpae*,
                                                                        p. 75

      Leaves and fruits not as above; sepals erect;
            branchlets often bearing spines   .   .   3. sect. *Abbreviatae*,
                                                                        p. 75

   Leaves not in fascicles terminating short shoots,
         though sometimes ± crowded towards the
         ends of lateral shoots:

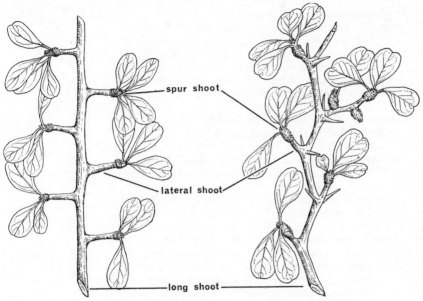

FIG. 11. Diagrammatic illustration of shoot types in *Terminalia*. Drawn by Mrs. M. E. Church.

Bark on young branches purple-black, peeling off in cylindric or hemicylindric papery flakes, leaving a reddish-brown or brown (later grey-brown) newly exposed surface	4. sect. *Psidioides*, p. 76
Bark on young branches not peeling as above	5. sect. *Platycarpae*, p. 77

Sect. 1. **Fatrea** (*Juss.*) *Exell* in Kirkia 7: 229 (1970)

Syn. *Fatrea* Juss. in Ann. Mus. Nat. Hist. Nat. Paris 5: 223 (1804)
[Sect. *Myrobalanus* sensu Engl. & Diels in E.M. 4: 9 (1900), *non* (Gaertn.) DC.]

Leaves small, spirally arranged or alternate, often on spur shoots. Inflorescences of axillary spikes often on spur shoots. Flowers all ♀. Fruit ellipsoid-ovoid, not or obscurely ridged.

One species in E. Africa (coastal) . . . 1. *T. boivinii*

Sect. 2. **Discocarpae** *Engl. & Diels* in E.M. 4: 26 (1900)

Leaves subcircular, long petiolate, spirally arranged on stunted spur shoots, precocious. Fruit subcircular; wings developed beyond the apical peg and the insertion of the stipe.

One species for the section . . . . 2. *T. orbicularis*

Sect. 3. **Abbreviatae** *Exell* in Bol. Soc. Brot., sér. 2, 42: 30 (1968)

Syn. Sect. *Platycarpae* Engl. & Diels in E.M. 4: 17 (1900), pro parte

Leaves in fascicles on short spur shoots; branchlets often bearing spines. Fruit 2-winged; wings not developed beyond the apical peg and the insertion of the stipe.

Long shoots straight or almost so:
  Lateral shoots inserted ± regularly at ± 4–6 cm.
    intervals along the long shoot, diverging
    almost at right-angles, ending in a spur
    shoot; spur shoots rarely borne on long
    shoots; spines, when present, only on long
    shoots; fruit elliptic-oblong, 4–6·5 cm. long,
    2–3 cm. wide   .   .   .   .   .   .   3. *T. prunioides*
  Lateral shoots scattered irregularly along the
    long shoots and diverging at an obtuse
    angle:
      Leaves more than 1 cm. long; lateral shoots
        terminating in a spine and bearing 1–3
        spur shoots: spur shoots and spines
        often borne directly on long shoots;
        fruit subcircular, 1·8–2·4 cm. long, 1·9–2
        cm. wide   .   .   .   .   .   .   4. *T. brevipes*
      Leaves not more than 1 cm. long; lateral
        shoots terminating in a spur shoot (rarely
        in a spine) and bearing at least 10 spur
        shoots; numerous spur shoots also borne
        on long shoots; spines never present on
        long shoots; fruit ovate-elliptic, 1·8–2 cm.
        long, 1·2–1·4 cm. wide   .   .   .   .   5. *T. parvula*
Long shoots zigzag:
  Lower receptacle and fruit glabrous; spines
    usually present at base of spur shoots:
      Leaves scarcely longer than broad, broadly
        obovate, light green, never yellowish
        green beneath; spines present at base of
        every spur shoot   .   .   .   .   .   6. *T. spinosa*
      Leaves much longer than broad, obovate-
        elliptic to narrowly obovate-elliptic,
        yellowish green beneath; spines sometimes
        present at the base of spur shoots   .   .   7. *T. stuhlmannii*
  Lower receptacle tomentose; fruit finely puber-
    ulous; spines absent   .   .   .   .   8. *T. polycarpa*

Sect. 4. **Psidioides** *Exell* in Bol. Soc. Brot., sér. 2, 42: 30 (1968)

SYN. Sect. *Platycarpae* Engl. & Diels in E.M. 4: 17 (1900), pro parte

   Leaves not in fascicles terminating short spur shoots. Bark on young
branchlets peeling off in cylindric or hemicylindric flakes, leaving a reddish
brown or brown (later grey-brown) newly exposed surface. Fruit 2-winged.

NOTE. Intermediates, probably of hybrid origin, are likely to be encountered between
  species of this section.

Leaves mostly sessile or subsessile, glabrous or
    nearly so; lower receptacle glabrous   .   .   9. *T. brachystemma*
Leaves petiolate:
  Leaves glabrous or nearly so:
    Lamina narrowly elliptic, up to 12·5 cm. long
      and 3·5 cm. wide; venation obscure   .   10. *T. kaiserana*
    Lamina oblong-lanceolate to obovate, more
      than 12 cm. long and 4 cm. wide; venation
      conspicuous   .   .   .   .   .   14. *T. sp. B*

Leaves silky pubescent to tomentose beneath
(sometimes glabrescent when very old);
upper receptacle densely hairy :
   Secondary leaf-nerves conspicuously raised
beneath; indumentum of lower leaf-
surface and lower receptacle coarsely
sericeous or velutinous; fruits 6–7·5 cm.
long :
      Leaves usually oblanceolate, less often
obovate or elliptic-oblanceolate, long-
cuneate (the actual base acute to
shortly rounded), mostly 2·5–5 cm.
wide, the reticulation (if visible between
hairs) not or scarcely raised   .   .   11. *T. trichopoda*
      Leaves elliptic or elliptic-obovate, with the
margin curved gradually to the obtuse
or rounded base, mostly 6·5–9 cm. wide,
the ultimate reticulation prominent
(T1)   .   .   .   .   .   .   13. *T. sp. A*
   Secondary leaf-nerves not raised beneath;
indumentum of lower leaf-surface and
lower receptacle finely sericeous; fruits
3–4 cm. long   .   .   .   .   12. *T. sericea*

Sect. 5. **Platycarpae** *Engl. & Diels* in E.M. 4 : 17 (1900), emend. Exell in
Bol. Soc. Brot., sér. 2, 42 : 31 (1968) & in Kirkia 7 : 239 (1970)

SYN. Sect. *Stenocarpae* Engl. & Diels in E.M. 4 : 11 (1900)

   Leaves spirally arranged, not borne in fascicles on short shoots; bark of
branchlets not peeling off in cylindrical or hemicylindrical flakes. Fruit
2-winged.

Branchlets becoming corky in second year, later
developing a thick layer of cork; fruits
oblong to oblong-elliptic, 3 times as long
as broad :
   Leaves sessile, or if shortly petiolate (petiole
up to 2 cm. long) then lamina usually
decurrent almost to base of petiole   .   .   15. *T. macroptera*
   Leaves with a distinct petiole 2–5 cm. long;
lamina not decurrent :
      Leaves usually 16–22 cm. long, 6–8 cm. wide,
venation on lower surface obscurely
reticulate, not prominent; fruit glabrous
(at least when mature), 5·7–8 cm. long,
2·6–3·5 cm. wide; flowers glabrous,
pubescent or tomentose   .   .   .   16. *T. laxiflora*
      Leaves 16–32 cm. long, 7–11 cm. wide, venation
on lower surface conspicuously reticulate,
prominent or subprominent; fruit densely
velutinous-tomentose even when mature,
6·5–11·5 cm. long, 2·8–5·5 cm. wide;
flowers always densely tomentose .   .   17. *T. mollis*
Branchlets remaining fibrous after several years
growth :
Fruit at least 3 times as long as broad, narrowly

oblong to oblong-elliptic, 5·5–7·5 cm. long, 1·5–2·2 cm. wide; leaves broadly oblanceolate-elliptic, elliptic or broadly oblong-lanceolate, sparsely to densely sericeous-tomentose on the prominent nerves beneath; flowers densely tomentose . . . . . 18. *T. glaucescens*

Fruit always much less than 3 times as long as broad:

Older branchlets with very prominent leafscars; apical bud conspicuous, silvery grey, ± 7 mm. long and wide; leaves elliptic to oblong-elliptic, generally rather rounded and somewhat asymmetric at the base and narrowed gradually towards the rounded to acute apex, usually somewhat glaucous beneath; flowers densely tomentose with spreading hairs . . 19. *T. stenostachya*

Older branchlets and apical buds not as above; leaves ± broadly obovate to elliptic-obovate, apex shortly acuminate or cuspidate, base cuneate:

Flowers and fruit glabrous . . . 20. *T. brownii*

Flowers densely tomentose; fruit minutely puberulous:

Tree up to 10 m. high; leaves rounded and shortly apiculate at the apex, densely grey pilose beneath and only tardily glabrescent, undersurface light brown . . . . . 21. *T. kilimandscharica*

Tree up to 39 m. high; leaves distinctly acuminate, pilose but soon glabrescent except for the venation, undersurface yellowish green to olive brown 22. *T. sambesiaca*

1. **T. boivinii** *Tul.* in Ann. Sci. Nat., sér. 4, 6: 95 (1856) as "*bovinii*"; Capuron, Combrét. Arbust. ou Arboresc. Madagasc.: 83, t. 19/1–4 (1967); Exell in Kirkia 7: 230 (1970). Type: NE. Madagascar, Vohémar, *Bernier* 266 (P, holo.)

Shrub or small tree up to 5 m. high; wood very hard; bark smooth, greyish brown; long shoots straight; lateral shoots terminating in a spur shoot and bearing 4–6 lateral spurs 2–5 mm. long; some spur shoots borne directly on the long shoots. Leaves alternate; lamina chartaceous, narrowly obovate-elliptic to obovate, 2·5–6 cm. long, 1–2·5 cm. wide, apex obtuse to rounded, base acute to cuneate, green and shiny above, somewhat brownish or yellow-green beneath, glabrous when mature; lateral nerves 6–7 pairs; petiole 1–5 mm. long. Inflorescences of lateral spikes usually ± 1·5 cm. long, often on spur shoots; peduncle glabrous. Flowers yellowish, nearly sessile, glabrous outside or almost so. Fruit, pale yellowish green, ellipsoid-ovoid, 8–12 mm. long, 4·5–6 mm. wide, glabrous, not ridged, very shortly stipitate.

KENYA. Mombasa or vicinity, Nov. 1884, *Wakefield*!; Kilifi District: Mida, Oct. 1930, *Donald* 96 in F.D. 2496! & Hadu [Adu], 4 Mar. 1959, *Moomaw* 1602!
TANGANYIKA. Tanga District: Kigombe beach, 12 July 1953, *Drummond & Hemsley* 3251!; Bagamoyo District: Kikoka Forest Reserve, 27 May 1964, *Semsei* 3725!; Rufiji District: Mafia I., Bweni–Ras Mkumbi, 12 Aug. 1937, *Greenway* 5062!

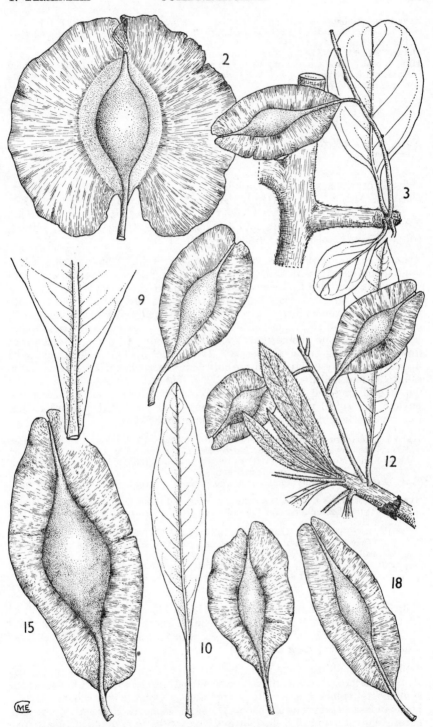

FIG. 12. Leaves and fruits, × 1, of *Terminalia* species numbered as in text—2, fruit of *T. orbicularis*; 3, fruiting branchlet of *T. prunioides*; 9, base of leaf and fruit of *T. brachystemma*; 10, leaf and fruit of *T. kaiserana*; 12, fruiting branchlet of *T. sericea*; 15, fruit of *T. macroptera*; 18, fruit of *T. glaucescens*. Drawn by Mrs. M. E. Church.

ZANZIBAR. Zanzibar I., Kiwengwa–Kinyasini, 29 Jan. 1929, *Greenway* 1246! & 1247! & Mazizini [Massazine], 14 June 1959, *Faulkner* 2255!; Pemba I., Ngugu, 15 Feb. 1929, *Greenway* 1447!

DISTR. **K**7; **T**3, 6, 8; **Z**; **P**; Mozambique and Madagascar

HAB. Coastal bushland and dry evergreen forest, often forming a thick cover near high water mark, growing on a wide variety of soils from sands and coral to heavy cracking clays; sea-level to 30 m.

SYN. [*T. fatraea* sensu P.O.A. C: 295 (1895); Engl. & Diels in E.M. 4: 9, t. 15/D (1900); T.T.C.L.: 143 (1949); U.O.P.Z.: 466 (1949); Griffiths in J.L.S. 55: 825, fig. 1 (1959); K.T.S.: 152 (1961), *non* (Poir.) DC.]

NOTE. This species differs from *T. fatraea*, a wholly Madagascan species with which it has been confused in the past, by having smaller fruits without longitudinal ridges.

2. **T. orbicularis** *Engl. & Diels* in E.M. 4: 26, t. 15/A (1900); Jex-Blake, Gard. E. Afr., ed. 3, t. 2 (1949); Griffiths in J.L.S. 55: 827, fig. 2 (1959); K.T.S.: 153 (1961). Types: Kenya, Machakos/Kitui/Teita Districts, Ndi–Ukambani, *Hildebrandt* 2607 (B, holo. †); Northern Frontier, Dandu, *Gillett* 12559 (K, neo.!, see Griffiths, *loc. cit.*: 829)

Large spreading shrub or small tree up to 8 m. high; bark smooth, grey; long shoots zigzag with lateral branches diverging at an acute angle and ending in a spur shoot. Leaves spirally arranged on the spur shoots; lamina suborbicular, up to 6·5 cm. long and wide, apex broadly and shortly acuminate, base rounded to subcordate, occasionally decurrent, glabrous; lateral nerves 3–6 pairs; petiole 2–6 cm. long. Inflorescences of lateral spikes borne on the spur shoots, 2·5–5 cm. long; peduncle densely tomentose; flowers appearing before the leaves. Lower receptacle densely tomentose; upper receptacle red and tomentose only at the base. Sepals cream internally, reflexed when mature. Fruit (fig. 12/2, p. 79) red, suborbicular in outline, 6·5–10 cm. long, 5·5–9·5 cm. wide; body finely pubescent; wings developed beyond the apical peg and the insertion of the stipe, the projections frequently overlapping; stipe up to 1 cm. long, pubescent.

KENYA. Northern Frontier Province: Dandu, 17 Mar. 1952, *Gillett* 12559! & Ndoto Mts., Latakwen, 17 Dec. 1958, *Newbould* 3257!; Machakos District: 16 km. on Mtito Andei–Mombasa road, 21 Dec. 1954, *Verdcourt* 1169!

DISTR. **K**1, 4, 7; Somali Republic and Ethiopia

HAB. Frequent to locally dominant in *Commiphora*, *Acacia* bushland; 500–1800 m.

SYN. *T. orbicularis* Engl. & Diels var. *macrocarpa* Engl. & Diels in E.M. 4: 26 (1900). Type: Ethiopia/Somali Republic (S.), Uebi [Webi] valley, *Brichetti* 232 (FI, holo.)
T. ruspolii Engl. & Diels in E.M. 4: 26, t. 15/B (1900). Types: Kenya, Tsavo [Tsawo], *Pospischil* (W, syn., B, isosyn. †) & Ethiopia, Uebi Scebeli [Ueb] valley, *Riva* 118 & 912 (FI, syn.)
T. praecox Engl. & Diels in E.M. 4: 27, t. 15/C (1900). Type: Kenya, Tana River District, Fullekullesat, *F. Thomas* 60 (B, holo. †)

3. **T. prunioides** *Laws.* in F.T.A. 2: 415 (1871); Engl. & Diels in E.M. 4: 22, t. 11/A (1900); T.S.K., ed. 2: 31 (1936); Griffiths in J.L.S. 55: 829, fig. 3 (1959); K.T.S.: 153, fig. 30/i (1961); Exell in Kirkia 7: 231 (1970). Types: Mozambique, Tete, *Kirk* (K, syn.!)

Shrub or small tree 7–15 m. high; bark grey, fissured; long shoots usually straight; lateral shoots ending in spur shoots, rarely with additional lateral spur shoots; spines occasionally present and only on the long shoots. Leaves borne in fascicles on the spur shoots; lamina chartaceous, broadly obovate to elliptic-obovate, up to 7·5 cm. long and 3 cm. wide, apex rounded, emarginate or mucronate, base obtuse to cuneate, usually densely pubescent when young, becoming glabrescent; lateral nerves 3–5 pairs, ± impressed above; petiole

0·5–1·5(–2·5) cm. long. Inflorescences of lateral spikes 5–8 cm. long; peduncle
1·7–2 cm. long, densely to sparsely pubescent. Flowers cream or white,
glabrous or nearly so. Fruit (fig. 12/3, p. 79) purplish brown or red, elliptic-
oblong in outline, 4–6·5 cm. long, 2–3 cm. wide, apex obtuse, deeply emarginate or mucronate; stipe up to 7 mm. long. Cotyledons 2, with petioles
3–4 mm. long, borne above soil-level.

KENYA. Teita District: 8 km. NW. of Tsavo R. on Mombasa road, 2 Feb. 1952, *Trapnell*
    2223 !; Tana River District: Karawa, 14 Oct. 1961, *Polhill & Paulo* 659 !; Lamu
    District: Pate [Patta] I., Feb. 1957, *Greenway & Rawlins* 8929 !
TANGANYIKA. Lushoto District: Mkomazi, 8 Jan. 1966, *Leippert* 6194 !; Tanga District:
    S. of Umba R. on Mombasa–Tanga road, 4 Mar. 1956, *Greenway* 8969 ! & Perani
    Forest, 12 Aug. 1953, *Drummond & Hemsley* 3706 !
DISTR. K1, 4, 7; T3; Mozambique, Zambia, Rhodesia, Botswana, Angola, South West
    Africa and South Africa (Transvaal)
HAB. *Acacia*, *Commiphora* and *Acacia*, *Combretum* bushland, coastal bushland and
    riverine thicket; 30–1400 m.

SYN.   *T. holstii* Engl. in Abh. Preuss. Akad. Wiss. 1894: 34 (1894) & P.O.A. C: 294
      (1895); T.T.C.L.: 145 (1949). Type: Tanganyika, Lushoto District, Bwiti
      [Buiti], *Holst* 2381 (B, holo. †, K, iso. !)*
    T. petersii Engl., P.O.A. C: 294 (1895). Type: Mozambique, Tete, *Peters* (B,
      holo. †)
    [*T. polycarpa* sensu Griffiths in J.L.S. 55: 829 (1959), pro parte quoad specim.
      Dale 745, *non* Engl. & Diels]

4. **T. brevipes** *Pampan.* in Bull. Soc. Bot. Ital. 1915: 16 (1915); Griffiths
in J.L.S. 55: 834, fig. 4 (1959); K.T.S.: 151, fig. 30/C (1961). Type: Somali
Republic (S.), Uebi Scebeli, between Muccoidère and Balaad, *Paoli* 1339
(FI, holo., K, photo. !)

Small tree or shrub, sometimes scandent, up to 10 m. high; bark rough,
thorny; branchlets fibrous, light to greyish brown; long shoots straight,
rather slender, spur shoots and spines occasionally present; lateral shoots
irregularly arranged, diverging at an obtuse angle, ending in a spine, bearing
1–4 spur shoots. Leaves borne on the spur shoots; lamina broadly oblanceolate to narrowly or broadly obovate, up to 6·5 cm. long and 3·5 cm. wide,
usually smaller, apex obtuse to rounded, sometimes emarginate, base acute
to cuneate, glabrous to sparsely pubescent beneath; lateral nerves 3–4 pairs;
petiole up to 1 cm. long. Inflorescences of slender spikes up to 7 cm. long,
borne on the spur shoots; peduncles sparsely pilose. Flowers white, shortly
pedicellate, glabrous outside. Fruit yellowish brown, subcircular, 1·8–2·4
cm. long, 1·9–2 cm. wide, glabrous; apical peg ± 0·5 mm. long; stipe ±
1–2 mm. long.

KENYA. Kilifi District: Galana R., N. of Malindi, July 1959, *Tweedie* 1866 !; Tana River
    District: Garissa, 4 Feb. 1956, *Greenway* 8863 ! & Bura, 12 Mar. 1963, *Thairu* 66 !
TANGANYIKA. Mbulu District: Tarangire, 7 Jan. 1959, *Mahinda* 439 !
DISTR. K1, 7; T2; Ethiopia and Somali Republic
HAB. Riverine forest and bushland, sometimes dominant, alluvial soils; 20–280 m.

SYN.   *T. balladellii* Chiov., Fl. Somala 2: 207 (1932). Type: Somali Republic (S.),
      Juba, Lak Dera, *Balladelli* 264 (FI, holo. !, K, photo. !)

5. **T. parvula** *Pampan.* in Bull. Soc. Bot. Ital. 1915: 17 (1915); Griffiths
in J.L.S. 55: 835, fig. 5 (1959); K.T.S.: 153, fig. 30/h (1961). Type: Somali
Republic (S.), Juba R., Revai–Sorori, *Paoli* 497 & Bardera–Mansur, *Paoli*
580 & near Lugh, *Paoli* 992a & b (all FI, syn., K, photo. !)

Small tree 3–5 m. high; bark black or grey, striate; branchlets becoming
fibrous, whitish to purplish grey; long shoots straight, bearing numerous

* The isotype at Kew gives the locality as Simbili, which is, however, in the same area.

small spur shoots; lateral shoots irregularly arranged, sometimes opposite, subopposite or in whorls of 3, diverging at an obtuse angle, ending in a spur shoot, rarely in a spine and bearing numerous spur shoots. Leaves borne on the spur shoots; lamina obovate or spathulate, up to 1 cm. long and 0·5 cm. wide, apex rounded, base cuneate, glabrous; lateral nerves 2–3 pairs, inconspicuous. Inflorescences of slender spikes up to 1·5 cm. long, borne on the spur shoots; peduncle glabrous. Flowers white, shortly pedicellate, glabrous. Fruit reddish, ovate-elliptic, 1·8–2·5 cm. long, 1·2–1·8 cm. wide, glabrous, reddish, apex rounded, emarginate, base cuneate to subtruncate; apical peg absent; stipe up to 0·5 cm. long.

KENYA. Teita District: Tsavo National Park, East, Voi Gate–Lugard Falls, 1 Jan. 1967, *Greenway & Kanuri* 12919 !; Kilifi District: Lali Hills, 15 Nov. 1967, *D. Wood* 1371 !; Tana River District: Galole, 17 Dec. 1964, *Gillett* 16379 !
DISTR. K1, 4, 7; Somali Republic
HAB. Occasional to common or even locally dominant in *Acacia, Commiphora* bushland; 60–870 m.

6. **T. spinosa** *Engl.*, P.O.A. C: 294 (1895); Engl. & Diels in E.M. 4: 25, t. 13/B (1900); T.T.C.L.: 146 (1949); F.P.S. 1: 208 (1950); I.T.U., ed. 2: 93 (1952); Griffiths in J.L.S. 55: 836, fig. 6 (1959); K.T.S.: 154, fig. 30/g (1961). Types: Kenya, Mombasa, *Wakefield* (B, syn. †, K, isosyn. !) & Tanganyika, Pangani, *Stuhlmann* I.355 (B, syn. †)

Tree up to 20 m. high; bark grey, longitudinally fissured; branchlets greyish or reddish brown; long shoots zigzag with occasional spur shoots; lateral shoots irregularly arranged, diverging at an obtuse angle, ending in a spur shoot and with 0–2 lateral spur shoots; spines 2(–3), frequently at base of spur shoots, up to 1·5(–2) cm. long. Leaves borne on the spur shoots; lamina broadly obovate, up to 4 cm. long and 3 cm. wide, apex rounded to subtruncate, emarginate, base cuneate, glabrous to sparsely pubescent; lateral nerves 4–5 pairs; petiole up to 5 mm. long. Inflorescences of spikes up to 6 cm. long, borne on the spur shoots; peduncle sparsely to densely pubescent. Flowers white or purplish, shortly pedicellate, glabrous outside. Fruit reddish purple, oblong-elliptic, 2–3 cm. long, 1–2 cm. wide, apex emarginate, glabrous; apical peg absent; stipe 3 mm. long.

UGANDA. Karamoja District: Lodoketemit, 29 Jan. 1959, *Kerfoot* 813 ! & Amudat, Jan. 1965, *Newbould* 6779 ! & Kokumongole [Kakumongole], 7 Jan. 1937, *A. S. Thomas* 2200 !
KENYA. Kitui District: Mutomo [Motomo] Hill, 20 Jan. 1942, *Bally* 1587 !; Kilifi District: Watamu–Mida Creek, 16 Oct. 1962, *Greenway* 10838 !; Tana River District: Kurawa, 25 Sept. 1961, *Polhill & Paulo* 560 !
TANGANYIKA. Pangani District: Mkwaja Ranch, 26 Nov. 1955, *Tanner* 2395 !; Morogoro District: Wami Flats, 9 Dec. 1933, *B. D. Burtt* 5047 ! & Morogoro Fuel Reserve, Nov. 1954, *Semsei* 1866 !
DISTR. U1; K1–4, 7; T2, 3, 6; Sudan and Somali Republic
HAB. *Acacia, Commiphora* and coastal bushland, less often in *Brachystegia* woodland and wooded grassland; 0–1770 m.

7. **T. stuhlmannii** *Engl.*, P.O.A. C: 294 (1895); Engl. & Diels in E.M. 4: 22, t. 10/A (1900); T.T.C.L.: 146 (1949); Griffiths in J.L.S. 55: 840, fig. 8 (1959); Exell in Kirkia 7: 232 (1970). Type: Tanganyika, Mpwapwa, *Stuhlmann* 204 (B, holo. †); Mozambique, Tete, near Boroma, *Menyharth* 766 (K, neo. !—Griffiths, *loc. cit.*: 842 (1959) has incorrectly cited this as a lectotype)

Small tree up to 12 m. high, sometimes a shrub; bark brownish or whitish grey; long shoots zigzag with occasional spur shoots; lateral shoots irregularly arranged, diverging at an obtuse angle, ending in a spur shoot, lateral spur shoots rare; spines up to 1·5(–2·5) cm. long sometimes present at the base of spur shoots. Leaves borne on the spur shoots; lamina obovate-elliptic to narrowly obovate-elliptic, up to 6 cm. long and 3 cm. wide, apex rounded to obtuse, base cuneate, coriaceous, nearly glabrous, glaucous green above, yellowish green beneath; lateral nerves 4–5 pairs; petiole up to 4 mm. long. Inflorescences up to 8 cm. long; peduncle glabrous. Flowers whitish, glabrous outside. Fruit purple-pink or brown, elliptic, 2·3–4 cm. long, 1·6–3 cm. wide, glabrous; apical peg absent; stipe 5–7 mm. long. Cotyledons 2, with petioles 2–4 mm. long, borne above soil-level.

TANGANYIKA. Mbulu District: Tarangire, 7 Jan. 1959, *Mahinda* 439!; Nzega District: Mpumbulya, 3 Apr. 1958, *Howard* 37!; Iringa, 4 Feb. 1962, *Polhill & Paulo* 1350!
DISTR. T1, 2, 4–7; Mozambique, Zambia, Rhodesia and Botswana
HAB. Deciduous woodland, bushland and thicket, wooded grassland, often on rather poorly drained soils; 840–1500 m.

8. **T. polycarpa** *Engl. & Diels* in E.M. 4: 24, t. 12/A (1900); Griffiths in J.L.S. 55: 843, fig. 9 (1959), excl. specim. *Dale* 745; K.T.S.: 153, fig. 30/e (1961). Type: Kenya, Northern Frontier Province, R. Daua, Beila, *Riva* 1466 [1620, 638] (FI, lecto., see Griffiths *loc. cit.*, K, photo.!)

Small tree 5–8 m. high; bark dark grey, smooth; branchlets fibrous, greyish brown to purplish grey. Long shoots zigzag, with occasional spur shoots; lateral shoots irregularly arranged, diverging at an obtuse angle, ending in a spur shoot and with up to 2 lateral spur shoots; spines absent. Leaves borne on the spur shoots; lamina obovate, up to 5·5 cm. long and 3·2 cm. wide (*fide* Griffiths), apex rounded, broadly emarginate, base acute to cuneate, glabrous to sericeous-pubescent; lateral nerves 4–5 pairs; petiole up to 10 mm. long. Inflorescences of spikes up to 7·5 cm. long, borne on the spur shoots; peduncle tomentose. Flowers cream; lower receptacle tomentose; upper receptacle sparsely pubescent. Fruit yellowish purple, elliptic-oblong, 2·5–3 cm. long, 1·2–1·5(–2) cm. wide, finely puberulous; apical peg absent; stipe ± 4 mm. long.

KENYA. Northern Frontier Province: 24 km. S. of Moyale, 2 Sept. 1953, *Bally* 9077! & Dandu, 22 Mar. 1952, *Gillett* 12610!
DISTR. K1; Ethiopia and Somali Republic
HAB. Semi-desert scrub; 670–870 m.

SYN. *T. somalensis* Engl. & Diels in E.M. 4: 24, t. 10/C (1900); Griffiths in J.L.S. 55: 845 (1959); K.T.S.: 154 (1961). Type: Ethiopia, Ogaden, *Bricchetti* 126 (FI, lecto., see Griffiths *loc. cit.*, K, photo.!)
    T. kelleri Engl. & Diels in E.M. 4: 24, t. 10/B (1900). Types: Ethiopia, Ogaden, Abdulla [Abdallah], *Keller* 222 & 225 (both Z, syn., K, photo.!)

NOTE. More material of this uncommon and rather variable tree is required.

9. **T. brachystemma** *Hiern*, Cat. Afr. Pl. Welw. 2: 340 (1898); Engl. & Diels in E.M. 4: 20, t. 9/C (1900); Griffiths in J.L.S. 55: 846, fig. 11 (1959), excl. specim. *Eggeling* 6089; Liben in F.C.B., Combr.: 93 (1968); Exell in Kirkia 7: 237 (1970). Type: Angola, Huila, Lopollo–Empalanco, *Welwitsch* 4287 (BM, lecto.!, COI, K!, LISU, P, isolecto.)

Small tree 5–8 m. high; bark grey, longitudinally fissured; branchlets pubescent or glabrous with purplish brown or purplish black bark peeling off in strips to reveal a light brown newly exposed surface. Leaves spirally arranged, sessile or subsessile; lamina broadly obovate to obovate-elliptic, 9–16 cm. long, 5–7 cm. wide, apex obtuse to rounded, sometimes shortly cuspidate, base cuneate to slightly amplexicaul, glabrous or sparsely pubescent or sparsely appressed pilose especially on the midrib; lateral nerves 10–12 pairs. Inflorescences of axillary spikes 7–11 cm. long, often in the axils of fallen leaves; peduncles glabrous or pubescent. Flowers white;

lower receptacle glabrous or sparsely to densely sericeous; upper receptacle glabrous or sparsely pilose at the base. Fruit (fig. 12/9, p. 79) reddish brown, elliptic to elliptic-oblong, 4–5·5 cm. long, 2·3–3 cm. wide, apex obtuse to rounded and emarginate, base cuneate; apical peg absent; stipe 5–7 mm. long.

subsp. **sessilifolia** (*R. E. Fries*) *Wickens* in K.B. 25: 184 (1971). Type: Zambia, Mbala District, Msisi, *Fries* 1328 (UPS, holo., K, photo.!)

Leaves pubescent, though hairs sometimes difficult to detect on old leaves; lamina slightly amplexicaul at the base and leaving a deltate leaf-scar. Peduncle tomentose. Lower receptacle sericeous-tomentose.

TANGANYIKA. Ufipa District: Lake Sundu, 9 Sept. 1958, *Richards* 10260!
DISTR. **T4**; Zambia
HAB. Wooded grassland; 1500 m.

SYN. *T. sessilifolia* R. E. Fries, Wiss. Ergebn. Schwed. Rhod.-Kongo-Exped. 1: 174 (1914); Griffiths in J.L.S. 55: 851, fig. 13 (1959); F.F.N.R.: 290 (1962)

NOTE. Subsp. *brachystemma*, which is widespread from Zaire to Mozambique and south to Angola, Botswana and the Transvaal, has glabrous leaves, peduncles and receptacles and rounded or elliptical leaf-scars. The records from Tanganyika are based on specimens such as *Eggeling* 6089 and *Burrows* 14, which seem more correctly referable to *T. kaiserana* or forms intermediate between *T. kaiserana* and *T. sericea*. Sessile leaves such as those of *Burrows* 14 do occur on juvenile shoots of other specimens of *T. kaiserana*, e.g. *Joseph* 4039 and *Haerdi* 28/2B, but in these cases the presence also of adult leaves excludes any doubt as to their identity.

10. **T. kaiserana** *F. Hoffm.*, Beitr. Kenntn. Fl. Centr.-Ost-Afr.: 26 (1889); P.O.A. C: 294 (1895); Engl. & Diels in E.M. 4: 19, t. 13/D (1900); T.T.C.L.: 145 (1949); Griffiths in J.L.S. 55: 849, fig. 12 (1959); Exell in Kirkia 7: 239 (1970). Type: Tanganyika, Tabora District, Igonda [Gonda], *Boehm* 141a (B, holo. †)

Small tree or shrub, up to 10 m. high; bark fissured; branchlets sericeous-tomentose to glabrous; bark purplish brown or purplish black, peeling off in strips to reveal a light brown newly exposed surface. Leaves spirally arranged; lamina elliptic-oblanceolate or narrowly elliptic, 7·5–12·5 cm. long, 2·5–3·5 cm. wide, apex acute or shortly acuminate, base cuneate to narrowly cuneate, sparsely sericeous when young but soon glabrescent; lateral nerves 10–16 pairs, slightly conspicuous above, rather inconspicuous beneath; petiole up to 2·5(–3·5) cm. long, rather slender. Inflorescences of axillary spikes up to 11 cm. long; peduncle usually glabrous. Flowers cream or white; upper receptacle glabrous or sparsely pubescent. Fruit (fig. 12/10, p. 79) purplish brown, broadly elliptic, 4·5–5·5 cm. long, 2·5–3 cm. wide, apex obtuse to rounded and emarginate, base acute to subtruncate, glabrous; apical peg absent; stipe 5–7 mm. long.

TANGANYIKA. Tabora District: Kaliua, 5 Nov. 1949, *Shabani* 56!; Ulanga District: Ifakara, 2 Feb. 1959, *Haerdi* 28/2B!; Mbeya District: Igawa–Mbeya road, 4 Feb. 1961, *Richards* 14233!
DISTR. **T1, 4, 6, 7**; Burundi, Zambia and Malawi
HAB. *Brachystegia* woodland and wooded grassland; 400–1200(–1800) m.

SYN. *T. holtzii* Diels in E.J. 39: 513 (1907); T.T.C.L.: 145 (1949); Griffiths in J.L.S. 55: 905 (1959). Type: Tanganyika, Bukoba, *Holtz* 1637 (B, holo. †)
[*T. brachystemma* sensu Griffiths in J.L.S. 55: 848 (1959), pro specim. *Eggeling* 6089, *non* Hiern]

NOTE. This species occurs in an easily recognizable form between Lakes Nyasa, Tanganyika and Rukwa and then extends around the central plateau, on one side up Lake Tanganyika to Burundi and eastwards to Mwanza and Shinyanga, on the other eastwards to the Ulanga and Morogoro Districts. Where it extends inward from this arc and also in the eastern parts of its range, it overlaps the range of *T. sericea* and intermediates of various sorts are common (see 12 × 10, p. 86).

11. **T. trichopoda** *Diels* in E.J. 39: 514 (1907); Griffiths in J.L.S. 55: 858 (1959); F.F.N.R.: 290 (1962); Exell in Kirkia 7: 236 (1970). Types: Rhodesia, Matopos, *Engler* 2847a & Umtali, *Engler* 3142 (both B, syn. †)

Small tree up to 9(–21) m. high; bark dark grey; branchlets with purplish black bark peeling off in strips to reveal a light brown newly exposed surface. Leaves spirally arranged, petiolate; lamina usually oblanceolate, less often obovate or elliptic-oblanceolate, up to 16(–18) cm. long and 5(–7) cm. wide, apex usually acuminate, long-cuneate, usually rather coarsely brown tomentose or brown pilose, rarely pubescent or ± sericeous in specimens tending towards *T. sericea*; lateral nerves 10–14 pairs, fairly conspicuous beneath, as is also the reticulation; petiole 15–20 mm. long. Inflorescences of axillary spikes 10–14 cm. long; peduncle 4–5 cm. long, brown tomentose. Flowers white; lower receptacles densely tomentose, not sericeous; upper receptacle less densely tomentose. Fruit yellowish to reddish brown, broadly to narrowly oblong-elliptic, up to 7·5 cm. long and 4·2 cm. wide, finely tomentose; stipe 15–20 mm. long.

TANGANYIKA. Handeni District: Kwamsisi, Dec. 1964, *Procter* 2835!; Iringa District: Great Ruaha R. 11 km. W. of Kidatu bridge, 16 July 1970, *Thulin & Mhoro* 457!; Songea, 13 June 1956, *Milne-Redhead & Taylor* 10700!
DISTR. **T**3, 6–8; Zambia, Malawi, Mozambique, Rhodesia and Botswana
HAB. Deciduous woodland and wooded grassland; 450–1000 m.

SYN. ? *T. poliotricha* Diels in E.J. 54: 342 (1917); T.T.C.L.: 144 (1949); Griffiths in J.L.S. 55: 906 (1959). Type: Tanganyika, Morogoro District, Mlali valley, *Holtz* in *Herb. For. Dar es Salaam* 3144 (B, holo. †)

NOTE. Occurs to the east of the central plateau and generally readily distinguishable from *T. sericea* by the prominent nerves and matted tomentum of the leaves and by the large fruits, but these features are variable and intermediates with *T. sericea* and perhaps *T. kaiserana* are likely to be encountered.

12. **T. sericea** *DC.*, Prodr. 3: 13 (1828); Laws. in F.T.A. 2: 416 (1871); Engl. & Diels in E.M. 4: 20, t. 13/C (1900); T.T.C.L.: 145 (1949); Griffiths in J.L.S. 55: 853, fig. 14 (1959); Liben in F.C.B., Combr.: 94 (1968); Exell in Kirkia 7: 234 (1970). Type: South Africa, Cape Province, Chue valley to Mashowing R., *Burchell* 2399 (G, holo., K, iso.!)

Small spreading tree 3–16 m. high; bark grey-brown or pale cream, longitudinally fissured; branchlets with purplish black bark peeling off in strips to reveal a light brown newly exposed surface; young shoots sericeous-tomentose. Leaves spirally arranged; lamina narrowly obovate-elliptic to narrowly elliptic, 5·5–12·5 cm. long, 1·5–4·5 cm. wide, apex acute to rounded, base cuneate, sericeous-pubescent to densely silvery sericeous-tomentose beneath, becoming somewhat glabrescent when old; lateral nerves 5–8 pairs, usually inconspicuous; petiole 2–10 mm. long. Inflorescences of lateral spikes 5–7·5 cm. long; peduncle 2·5–3 cm. long, densely sericeous. Flowers greenish white, sericeous-tomentose outside. Fruit (fig. 12/12, p. 79) pinkish to purplish brown, broadly elliptic, 3–4 cm. long, 1·7–2·5 cm. wide, apex obtuse to rounded and usually emarginate, base obtuse to subtruncate, finely tomentose; stipe 5–7 mm. long. Cotyledons 2, petioles 1·5–2 cm. long, arising below soil-level.

TANGANYIKA. Tabora District: 72 km. N. of Tabora, 8 Dec. 1926, *Wallace* 1!; Dodoma District: Kazikazi, 20 Nov. 1932, *B. D. Burtt* 4592!; Tunduru, *Gillman* 1040!
DISTR. **T**1–8; Zaire, Mozambique, Malawi, Zambia, Rhodesia, Botswana, Angola, South Africa (Natal, Transvaal) and South West Africa
HAB. *Brachystegia* woodland and wooded grassland; 450–1300 m.

SYN. *T. angolensis* O. Hoffm. in Linnaea 43: 131 (1881). Type: Angola, Malange, collector unrecorded (B, holo. †)

T. angolensis Ficalho [in Bol. Soc. Geogr. Lisb. 2: 708 (1882), *nom. nud.*], Pl. Ut.
    Afr. Port.: 182 (1884), *non* O. Hoffm. (1881), *nom. illegit.* Type: Angola,
    Huila, *Welwitsch* 4343 (BM, lecto. !)

T. fischeri Engl., P.O.A. C: 294 (1895). Types: Tanganyika, Dodoma District,
    Saranda, *Fischer* 258 & 259 (B, syn. †)

T. nyassensis Engl., P.O.A. C: 294 (1895). Type: Malawi, Shire Highlands,
    Buchanan 1891 (B, holo. †, BM, K, iso. !)

T. brosigiana Engl. & Diels in N.B.G.B. 2: 191 (1898). Types: Tanganyika,
    near Morogoro & Kilosa District, Kimamba, *Brosig* (B, syn. †)

T. sericea DC. var. *angolensis* Hiern, Cat. Afr. Pl. Welw. 2: 338 (1898). Type: as
    T. angolensis Ficalho

T. sericea DC. var. *huillensis* Hiern, Cat. Afr. Pl. Welw. 2: 339 (1898). Type:
    Angola, Huila, Lopolo–Nene, *Welwitsch* 4294 (BM, lecto., COI, K !, LISU,
    isolecto.)

T. bubu De Wild. & Ledoux in De Wild., Contr. Fl. Kat., Suppl. 2: 84 (1929).
    Type: Zaire, Bas-Katanga, *Delevoy* 144 & 354 (BR, syn.)

NOTE. *T. sericea* is the commonly occurring species of the section on the central plateau
    of Tanganyika and easily recognized in **T5** and adjacent parts of the surrounding
    provinces, but where it extends further west into the Tabora District, southwards
    towards Zambia and east into **T6** and **T8** intermediates with *T. kaiserana* and perhaps
    T. trichopoda are likely to be encountered, see below.

12 × 10. **T. sericea × kaiserana**

Hairs on leaves often more persistent than in *T. kaiserana*, longer and less
obviously sericeous than in *T. sericea*. Indumentum developed to varying
degrees on the peduncle, rhachis, lower and upper receptacles, varying from
the typical situation in *T. kaiserana* where all these parts are glabrous to that
in *T. sericea* where all are densely hairy.

TANGANYIKA. Tabora District: Kaliua, 25 Oct. 1949, *Shabani*! & Tabora, 28 Sept.
    1961, *Boaler* 336!; Mpanda District: Lake Katavi, 16 Oct. 1965, *Richards* 20548!;
    Mbeya District: Igawa–Mbeya road, 4 Feb. 1961, *Richards* 14233!
DISTR. **T4, 6, 7, ? 8**; not recorded elsewhere

13. **T. sp. A**

Tree 10–13 m. high; bark grey, longitudinally fissured; branchlets with
purplish black bark peeling off in strips to reveal a light brown newly exposed
surface; young shoots densely yellowish brown pubescent. Leaves spirally
arranged; lamina elliptic to obovate-elliptic, 15–20 cm. long, 6·5–9 cm. wide,
apex acute to obtuse, base obtuse to rounded, sparsely pubescent above, grey
to yellowish brown pubescent on the venation beneath; lateral nerves 10–14
pairs, conspicuous above, prominent beneath, venation conspicuous beneath;
petiole 2–3·5 cm. long, pubescent. Inflorescence of lateral spikes 8–10 cm.
long; peduncle 2·5–3 cm. long, densely pubescent. Flowers pale yellow,
fragrant, densely sericeous-tomentose outside. Fruit yellowish to reddish
brown, oblong-elliptic, 7–7·5 cm. long, 3·5 cm. wide, apex obtuse, base
cuneate, shortly pubescent; stipe 10 mm. long.

TANGANYIKA. Mwanza District: near Minziwera, 30 Mar. 1937, *B. D. Burtt* 6465!;
    Musoma District: Lupa, 4 June 1959, *Tanner* 4316! & Nyambono, 6 June 1959,
    Tanner 4333!
DISTR. **T1**; not known elsewhere
HAB. Wooded grassland and riverine bushland; 1200–1440 m.

NOTE. This species appears to be related to *T. erici-rosenii* R. E. Fries from Zambia,
    from which it is readily distinguished by the larger fruit. The species also superficially
    resembles *T. stenostachya* and *T. mollis*, from which it can be positively identified by
    the purplish black bark peeling off in strips.

14. **T. sp. B**

Tree 7–12 m. high; branchlets with purplish black bark peeling off in
strips to reveal a light brown newly exposed surface; young shoots lightly

pubescent. Leaves spirally arranged; lamina oblong-lanceolate to obovate, 12–17 cm. long, 4–7 cm. wide, apex acute to obtuse, base cuneate, glabrous to sparsely pubescent; lateral nerves 10–16 pairs, conspicuous above, subprominent beneath; petiole up to 2·2 cm. long, lightly to densely hairy. Inflorescence 8–10 cm. long; peduncle 2–3 cm. long, pubescent; bracts lanceolate, 8–15 mm. long, persisting until flowers fully developed. Flowers yellow green, fragrant; lower receptacle densely hairy, upper receptacle less hairy. Fruit reddish brown, broadly ovate to oblong-ovate, 5–8 cm. long, 3·2–4·5 cm. wide, apex rounded to emarginate, base cuneate, pubescent when young, glabrous when mature; stipe 5–8 mm. long.

TANGANYIKA. Mwanza District: Bwiru, 12 Oct. 1952, *Tanner* 1057! & Bukumbi, Ngeleka, 24 Feb. 1953, *Tanner* 1244!; Kondoa District: Kolo, 12 Jan. 1962, *Polhill & Paulo* 1149!
DISTR. T1, 5, 7; not known elsewhere
HAB. Wooded grassland; 1140–1515 m.

NOTE. Probably related to 10, *T. kaiserana* from which it is readily distinguished by the larger leaves with a more pronounced venation and by the larger fruit.

15. **T. macroptera** *Guill. & Perr.*, Fl. Seneg. Tent. 1: 276, t. 63 (1832); Laws. in F.T.A. 2: 416 (1871); Engl. & Diels in E.M. 4: 11, fig. 2/A (1900); Burtt Davy, Check-lists Brit. Emp. 1, Uganda: 37 (1935); F.P.S. 1: 208 (1950); I.T.U., ed. 2: 90 (1952); F.W.T.A., ed. 2, 1: 279 (1954); Griffiths in J.L.S. 55: 864, fig. 17 (1959); Liben in F.C.B., Combr.: 97 (1968). Type: Gambia, Albreda, *Perrottet* (P, holo.)

Small tree up to 13 m. high; bark dark grey to black, deeply fissured; branchlets smooth, glabrous, purplish black, becoming light brown and corky with age. Leaves spirally arranged, sessile or almost so; lamina broadly oblanceolate to narrowly or broadly elliptic-oblanceolate, up to 37 cm. long and 17 cm. wide, apex rounded to obtuse or cuspidate or shortly cuneate, base cuneate and decurrent, glabrous or nearly so, margin undulate; lateral nerves 12–20 pairs; petiole, if present, up to 2 cm. long, often winged or partially winged. Inflorescences of lateral spikes up to 22 cm. long; peduncle up to 4 cm. long, glabrous or very sparsely pilose. Flowers white, glabrous outside. Fruit (fig. 12/15, p. 79) reddish brown, oblong-elliptic, 8–11 cm. long, 3·5–8 cm. wide, apex rounded to obtuse, emarginate or with apical peg ± 1 mm. long, glabrous; stipe 6–7 mm. long.

UGANDA. W. Nile District: Ladonga, Feb. 1934, *Eggeling* 1513!; Acholi, *Dawe* 857!; Teso District: Serere, July 1926, *Maitland* 1273!
DISTR. U1, 3; Gambia eastwards through Cameroun to the Sudan
HAB. Open woodland and wooded grassland; 770–1400 m.

SYN. *T. dawei* Rolfe in J.L.S. 37: 516 (1906); Burtt Davy, Check-lists Brit. Emp. 1, Uganda: 37 (1935); I.T.U., ed. 1: 47 (1940). Type: Uganda, Acholi, *Dawe* 865 (K, holo.!)
T. sp. sensu Burtt Davy, Check-lists Brit. Emp. 1, Uganda: 37 (1935), pro specim. *Chandler* 624! & *F.D.* 1443!

16. **T. laxiflora** *Engl. & Diels* in E.M. 4: 12, fig. 2/B (1900); F.P.S. 1: 208 (1950); F.W.T.A., ed. 2, 1: 279 (1954); Griffiths in J.L.S. 55: 868, fig. 18 (1959); Liben in F.C.B., Combr.: 98 (1968). Types: Sudan, Bahr el Ghazal, Tonj [Seriba Ghattas], *Schweinfurth* 1336 & 2088 (B, syn. †, K, isosyn.!) & 1343 & 1869 (B, syn. †)

Small tree up to 10 m. high; bark dark grey, deeply fissured. Leaves spirally arranged; lamina elliptic or oblong-elliptic or broadly oblanceolate, (13·5–)16–22(–30) cm. long, 4·5–9·5 cm. wide, apex rounded to obtuse or shortly acuminate, base obtuse to cuneate, glabrous or rarely tomentose beneath; lateral nerves 10–15 pairs; petiole 2·8–3·4 cm. long. Inflorescences of lateral spikes 9–12·5 cm. long; peduncle 1·7–2·2 cm. long, glabrous to densely tomentose. Flowers white; lower receptacle glabrous to densely tomentose; upper receptacle glabrous or sparsely pilose at the base. Fruit yellowish brown, oblong to oblong-elliptic, 5·7–8 cm. long, 2·6–3·5 cm. wide, apex obtuse to rounded, somewhat emarginate; apical peg often present; stipe 1–1·2 cm. long.

UGANDA. W. Nile District: Leya R. watershed, 25 Mar. 1945, *Greenway & Eggeling* 7249!; Karamoja District: Mt. Debasien, Jan. 1936, *Eggeling* 2774! & Napak, June 1950, *Eggeling* 5980!
DISTR. U1; Portuguese Guinea to Cameroun and eastwards to eastern Zaire, Sudan, and Ethiopia
HAB. Wooded grassland; 1830–2200 m.

SYN.   *T. schweinfurthii* Engl. & Diels in E.M. 4: 12 (1900); F.P.S. 1: 210 (1950). Types: Sudan/Ethiopia border, Matamma, *Steudner* 195 & 205 (both B, syn. †, K, isosyn.!) & *Schweinfurth* 2120 (B, syn. †, K, isosyn.!) & 2121 (B, syn. †)
       T. sp. aff. T. schweinfurthii Engl. & Diels sensu I.T.U., ed. 2: 94 (1952)

17. **T. mollis** *Laws.* in F.T.A. 2: 417 (1871); F.P.S. 1: 210 (1950); F.W. T.A., ed. 2, 1: 279 (1954); Griffiths in J.L.S. 55: 871, fig. 19 (1959); K.T.S.: 152, fig. 30/f (1961); Liben in F.C.B., Combr.: 98 (1968); Exell in Kirkia 7: 240 (1970), *non* Teysm. (1866), *nom. nud.*, *nec* v. Sloot. (1924), *nec* (Presl) Vidal (1885). Type: Sudan, without precise locality, *Petherick* (K, holo.!)

Small tree 5–13(–20) m. high; bark blackish grey, fissured; branchlets densely tomentose, with grey-brown to dark grey-brown bark becoming softly and thickly corky with age. Leaves spirally arranged, petiolate; lamina subcoriaceous, narrowly elliptic to elliptic or obovate-oblong, 16–37 cm. long, 7–19 cm. wide, apex obtuse to rounded or occasionally acute, base obtuse to rounded or occasionally subcordate, densely tomentose, brownish when dried; lateral nerves (7–)15–18 pairs, prominent beneath; petiole 3·5–5 cm. long, stout. Inflorescences of axillary spikes 8–17 cm. long; peduncles 1–2 cm. long, densely tomentose. Flowers cream, greenish white or pinkish, strong smelling, tomentose. Fruit pale yellowish green, oblong to broadly elliptic, 6·5–12 cm. long, 2·5–5·5 cm. wide, apex obtuse to rounded, emarginate, apical peg sometimes present, base acute to subtruncate, densely velutinous; stipe 5–7 mm. long. Cotyledons 2, with petioles 2–2·5 cm. long, arising below soil-level.

UGANDA. W. Nile District: Koboko, Feb. 1934, *Eggeling* 1526!; Acholi District: Gulu, Mar. 1934, *Tothill* 2493!; Mengo District: Nakasongola, 7 Sept. 1955, *Langdale-Brown* 1490!
KENYA. Trans-Nzoia District: Kitale, July 1933, *Dale* in *F.D.* 3093! & 5 Mar. 1953, *Bogdan* 3656!; N. Kavirondo District: Bungoma, 25 July 1951, *Greenway & Doughty* 8530!
TANGANYIKA. Ngara District: Keza, 25 Aug. 1960, *Tanner* 5111!; Mpanda District: Mwesi, Sept. 1961, *Procter* 1920!; Dodoma District: 55 km. on Itigi–Chunya road, 24 Apr. 1964, *Greenway & Polhill* 11701!
DISTR. U1, 3, 4; K3, 5; T1, 4, 5, 7, 8; W. Africa from Ghana to Cameroun and eastwards to the Sudan, also Zaire, Zambia and Angola
HAB. Wooded grassland, often co-dominant, and *Brachystegia* woodland; 930–2170 m.

SYN.   [*T. macroptera* sensu Oliv. in Trans. Linn. Soc. 29: 72 (1873), *non* Guill. & Perr.]
       T. torulosa F. Hoffm., Beitr. Kenntn. Fl. Centr.-Ost-Afr.: 27 (1889); P.O.A. C: 294 (1895); Engl. & Diels in E.M. 4: 15, fig. 5/A (1900), excl. specim. *Kirk*; Burtt Davy, Check-lists Brit. Emp. 1, Uganda: 37 (1935); T.S.K., ed. 2: 31 (1936); I.T.U., ed. 2: 93 (1952). Type: Tanganyika, Tabora District, Igonda [Gonda], *Boehm* 142A (B, holo. †)
       T. spekei Rolfe in J.L.S. 37: 516 (1906); Burtt Davy, Check-lists Brit. Emp. 1, Uganda: 37 (1935); I.T.U., ed. 1: 47 (1940). Types: Uganda, W. Nile District, Madi, *Grant* 643 & Acholi, *Dawe* 858 (both K, syn.!)

T. suberosa R. E. Fries, Wiss. Ergebn. Schwed. Rhod.-Kongo-Exped.: 172, t. 4 (1914). Types: Zambia, Chirukuta, *Fries* 264 & 264a (both UPS, syn., K, photo.!)

T. mildbraedii Mildbr., Z.A.E.: 581 (1914); T.T.C.L.: 143 (1949); Griffiths in J.L.S. 55: 905 (1959). Type: Tanganyika, Bukoba District, Kagera, *Mildbraed* 307 (B, holo. †)

T. sp. sensu Burtt Davy, Check-lists Brit. Emp. 1, Uganda: 37 (1935), pro specim. *Maitland* 1163! & *F.D.* 1451!

T. sp. cf. T. mollis sensu T.T.C.L.: 144 (1949)

NOTE. Several specimens, including *Newbould & Jefford* 2769! from Mahali Mts., *Verdcourt* 2820! from Kigoma and *Procter* 2263! from N. of Chunya, are apparently glabrescent forms of *T. mollis*, more material of which is required.

18. **T. glaucescens** *Benth.* in Hook., Niger Fl.: 336 (1849); Laws. in F.T.A. 2: 416 (1871); Engl. & Diels in E.M. 4: 13, fig. 4/B (1900); F.W.T.A., ed. 2, 1: 279, fig. 105/E (1954); Griffiths in J.L.S. 55: 879, fig. 21 (1959); Liben in F.C.B., Combr.: 100 (1968). Type: N. Nigeria, Lokoja, *Vogel* 178 (K, holo.!)

Small tree 7–13 m. high; bark light grey to black, deeply fissured; branchlets light or dark reddish brown. Leaves spirally arranged; lamina broadly oblanceolate-elliptic to elliptic or broadly oblong-lanceolate, 15–24(–27·5) cm. long, 5–11 cm. wide, apex obtuse to acute or shortly acuminate, base acute to rounded, densely tomentose when young, becoming densely sericeous-tomentose on the main nerves and sparsely pubescent on the tertiary nerves beneath; lateral nerves 9–16 pairs; petiole 2·5–4 cm. long. Inflorescences of axillary spikes 7–16 cm. long; peduncle 1·5–5 cm. long, densely pubescent. Flowers white or cream with a strong smell of carrion, densely sericeous-tomentose. Fruit (fig. 12/18, p. 79) brown, narrowly oblong to oblong-elliptic, 5·5–8 cm. long, 2·5–3·5 cm. wide, apex obtuse to truncate, usually emarginate, apical peg sometimes present, base obtuse to rounded, finely tomentose; stipe 5–10 mm. long.

UGANDA. W. Nile District: Amua, Sept. 1937, *Eggeling* 3424!; Teso District: Kagaa June 1936, *Eggeling* 3029!; Mengo District: Lwamatuka, 15 Sept. 1954, *Langdale-Brown* 1298!

TANGANYIKA. Bukoba District: Mubunda, Aug. 1957, *Procter* 685!

DISTR. **U**1–4; **T**1; in W. Africa from Guinée to Cameroun and east to Zaire, Sudan and Ethiopia

HAB. Wooded grassland, frequently dominant; 1000–1800 m.

SYN. *T. velutina* Rolfe in J.L.S. 37: 517 (1906); I.T.U., ed. 1: 48 (1940) & ed. 2: 93 (1952). Types: Uganda, Bunyoro District, Nkusi R., *Dawe* 697 & Busoga District, without precise locality, *E. Brown* 260 (both K, syn.!)

[*T. avicennioides* sensu Burtt Davy, Check-lists Brit. Emp. 1, Uganda: 37 (1935), non Guill. & Perr. (1832)]

T. sp. (*velutina* Rolfe?) sensu Burtt Davy, Check-Lists Brit. Emp. 1, Uganda: 37 (1935)

T. sp. near T. avicennioides sensu Burtt Davy, Check-lists Brit. Emp. 1, Uganda: 37 (1935)

T. sp. near T. schweinfurthii sensu Burtt Davy, Check-lists Brit. Emp. 1, Uganda: 37 (1935)

T. sp. sensu Burtt Davy, Check-lists Brit. Emp. 1, Uganda: 37 (1935), pro specim. *F.D.* 1459!

19. **T. stenostachya** *Engl. & Diels* in E.M. 4: 16, fig. 7/B (1900); T.T.C.L.: 144 (1949); Griffiths in J.L.S. 55: 888, fig. 24 (1959); Liben in F.C.B., Combr.: 101 (1968); Exell in Kirkia 7: 241 (1970). Type: Malawi, without locality, *Buchanan* 352 (BM, lecto.!, see Griffiths, *loc. cit.*)

Small tree 5–12(–20) m. high or shrub 4 m. high; bark dark grey to black, rough, lattice-like; branches densely tomentose when young, becoming glabrous with greyish brown fibrous bark and prominent leaf-scars. Leaves spirally arranged, petiolate; lamina elliptic to oblong-elliptic, 11–20 cm.

long, 4·5–11·5 cm. wide, apex acute to rounded, base obtuse to rounded or subcordate, coriaceous, tomentose when young, ± glabrescent; petiole 1·5–5 cm. long. Inflorescences of axillary spikes 10–16 cm. long; peduncle 2·5–4 cm. long, tomentose. Flowers cream, strong smelling; lower receptacle tomentose; upper receptacle sparsely pubescent. Fruit dark red or crimson, oblong to narrowly elliptic, (3–)4–6 cm. long, 2·4–3·5 cm. wide, apex obtuse to rounded, base obtuse to truncate, minutely puberulous; stipe 5–6 mm. long.

TANGANYIKA. Tabora District: Kigwa, *Howard* 4!; Mpanda District: Chipangati [Kipangati]–Sala, 3 Dec. 1950, *Bullock* 3511!; Chunya District: 16 km. on Chunya–Mbeya road, 19 Mar. 1965, *Richards* 19755!
DISTR. T(?1), 3–8; Zaire, Zambia, Rhodesia, Malawi and Mozambique
HAB. *Brachystegia* woodland and wooded grassland; 450–1400 m.

SYN.   ? *T. splendida* Engl. & Diels in E.M. 4: 16, t. 7/C (1900); T.T.C.L.: 144 (1949); Griffiths in J.L.S. 55: 906 (1959). Types: Tanganyika, Shinyanga District, Usiha area, *Fischer* s.n. & 256 (both B, syn. †)
       T. rhodesica R. E. Fries in Wiss. Ergebn. Schwed. Rhod.-Kongo-Exped. 1: 172, t. 13 (1914). Types: Zambia, Victoria Falls, *Fries* 78 & 78a (UPS, syn., K, photo.!)
       ? *T. dielsii* Engl., V.E. 3(2): 720 (1921); T.T.C.L.: 143 (1949). Type: Tanganyika, Morogoro District, Mlali valley, collector not cited

NOTE.   *T. stenostachya* is separated from *T. mollis* on the absence of development of a thick corky bark on the branchlets of the former. Young specimens or where insufficient length of branchlet has been collected are exceedingly difficult to separate.
        In the absence of type specimens or authenticated material the identity of both *T. splendida* and *T. dielsii* must remain uncertain.

20. **T. brownii** *Fresen.* in Mus. Senckenb. 2: 152, t. 9/1 (1837); Laws. in F.T.A. 2: 415 (1871); Engl. & Diels in E.M. 4: 17, fig. 8/A (1900); T.S.K., ed. 2: 31 (1936); F.P.S. 1: 210, fig. 115 (1950); I.T.U., ed. 2: 90 (1952); F.W.T.A., ed. 2, 1: 279 (1954); Griffiths in J.L.S. 55: 893, fig. 26 (1959); K.T.S.: 151, fig. 30/b (1961); Liben in F.C.B., Combr.: 100 (1968). Type: N. Ethiopia, Temben–Simen, *Rueppell* (FR, holo.)

Tree 7–13(–25) m. high; bark greyish, longitudinally fissured; branchlets grey, fibrous; young shoots tomentose, becoming glabrous. Leaves spirally arranged; lamina elliptic-obovate to broadly obovate, 6–16 cm. long, 3–8 cm. wide, apex obtuse to shortly acuminate, base acute to obtuse, sericeous-tomentose when young, becoming pilose beneath; lateral nerves 7–11 pairs; petiole 1·5–2(–4) cm. long. Inflorescences of axillary spikes up to 12 cm. long; peduncle 1·5–2 cm. long, tomentose. Flowers white or cream, fragrant, glabrous outside. Fruit reddish purple, broadly elliptic to ovate, 3·5–5·5 cm. long, 2·2–4 cm. wide, apex obtuse to rounded, emarginate, base acute to obtuse, glabrous, purplish red to brown; stipe 5–7 mm. long.

UGANDA. Karamoja District: Mt. Moroto, Apr. 1964, *J. Wilson* 1687!; Toro District: Sempaya R., 5 Dec. 1962, *Styles* 250!; Mbale District: Buligenyi, 29 Aug. 1932, *A. S. Thomas* 393!
KENYA. Northern Frontier Province: Ngare Narok, 27 Apr. 1944, *J. Bally* 56!; Baringo District: Marigat–Kabarnet, 30 Oct. 1964, *Leippert* 5256!; Machakos District: Lukenya, 17 Aug. 1958, *Napper* 751!
TANGANYIKA. Mbulu District: Minjingu, 4 Jan. 1962, *Polhill & Paulo* 1047!; Moshi District: Himo road, 14 June 1954, *Hughes* 198!; Kilosa, Nov. 1952, *Semsei* 1047!
DISTR. U1–3; K1–7; T2, 5, 6; Nigeria, eastern Zaire, Sudan, Ethiopia and the Somali Republic
HAB. Deciduous woodland, bushland and wooded grassland, riverine forest; 730–2000 m.

SYN.   *T. brownii* Fresen. var. *albertensis* Bagshawe & Bak. f. in J.B. 46: 7 (1908); Burtt Davy, Check-lists Brit. Emp. 1, Uganda: 37 (1935). Types: Uganda, Toro District, Semliki R., *Bagshawe* 1291 (BM, syn.!) & *Dawe* 634 (BM, syn., K, isosyn.!) & Mizizi R., *Bagshawe* 1319 (BM, syn.!)

T. semlikiensis De Wild., Pl. Bequaert. 4: 346 (1928). Type: Zaire, Semliki valley, Kasonsero, *Bequaert* 5050 (BR, holo., K, photo.!)

T. sp. sensu Burtt Davy, Check-lists Brit. Emp. 1, Uganda: 37 (1935), pro specim. *A. S. Thomas* 393!

21. **T. kilimandscharica** *Engl.*, P.O.A. C: 294 (1895); Engl. & Diels in E.M. 4: 19, fig. 9/B (1900); T.T.C.L.: 145 (1949); Griffiths in J.L.S. 55: 987, fig. 27 (1959), pro parte, excl. syn. *T. sambesiaca* & *T. aemula*; K.T.S.: 152, fig. 30/a (1961), pro parte, excl. specim. *Elliott* 1496. Type: Tanganyika, Kilimanjaro, *Johnston* (B, holo. †, BM, K, iso.!)

Small tree to 10(–13) m. high, or sometimes a shrub; bark grey-brown, deeply fissured; branchlets grey-brown, fibrous; young shoots tomentose, becoming glabrous. Leaves spirally arranged, petiolate; lamina obovate-elliptic to elliptic, up to 8·5(–11) cm. long and 5·5(–6·5) cm. wide, usually smaller, apex rounded, very shortly acuminate, base truncate to shortly attenuate, pinkish brown and lanate beneath, indumentum generally persisting; lateral nerves 7–9 pairs, somewhat prominent beneath; petiole 1–3 cm. long. Inflorescences of axillary or occasionally terminal spikes up to 8 cm. long; peduncle 1·5–3 cm. long, tomentose. Flowers white to cream, fragrant; lower receptacle tomentose; upper receptacle pilose at the base, almost glabrous towards the apex. Fruit brownish red, elliptic to oblong-elliptic, 7–11 cm. long, 4–6·5 cm. wide, apex acute to obtuse, sometimes emarginate, base narrowed into the stipe, puberulous to almost glabrous, but never entirely so; stipe 3–5 mm. long.

KENYA. Fort Hall District: 45 km. on Thika–Garissa road, 24 May 1957, *Verdcourt* 1782!; Machakos District: Kibwezi, 28 Nov. 1910, *Scheffler* 499!; Masai District: Garabani Hill, 6 Mar. 1940, *V. G. van Someren* 11!
TANGANYIKA. Masai District: Eluanata Farm, 3 May 1965, *Leippert* 5709!; Moshi District: near Himo, 14 Feb. 1936, *R. M. Davies* 1240! & near Moshi, 20 Dec. 1951, *McCoy-Hill* 14!
DISTR. **K**4, 6, 7; **T**2, 3, 5; not known elsewhere
HAB. Deciduous woodland, bushland and wooded grassland; 300–1700 m.

SYN. ? *T. canescens* Engl., P.O.A. C: 294 (1895); Engl. & Diels in E.M. 4: 18, t. 8/C (1900); T.T.C.L.: 145 (1949); Griffiths in J.L.S. 55: 904 (1959). Type: Tanganyika, Kondoa District, Irangi, *Stuhlmann* 4285 (B, holo.†)
T. sp. sensu T.S.K., ed. 2: 31 (1936)
[*T. brownii* sensu T.T.C.L. 144 (1949), pro parte quoad specim. *Greenway* 4453 & 4492, *non* Fresen.]
[*T. hildebrandtii* sensu T.T.C.L.: 145 (1949), ? sensu Engl., see note under *T. sambesiaca*]

NOTE. In the absence of type specimens or authenticated material the identity of *T. canescens* must remain uncertain.

22. **T. sambesiaca** *Engl. & Diels* in E.M. 4: 13, fig. 4/A (1900); Exell in Kirkia 7: 243 (1970). Type: Mozambique, Tete, near Boroma, *Menyharth* 613 (Z, holo., K, photo.!)

Tall tree up to 39 m. high; bark greyish, smooth to slightly rough and fissured; young branchlets tomentose, becoming glabrescent, with fibrous bark. Leaves spirally arranged, petiolate; lamina elliptic to broadly elliptic or obovate-elliptic, up to 18 cm. long and 13 cm. wide, apex rounded and acuminate, margin sometimes crenulate, base cuneate to obtuse or rounded, pubescent to pilose, especially on the nerves and reticulation beneath, often becoming glabrescent; lateral nerves 8–11 pairs, rather prominent beneath; petiole up to 4 cm. long, tomentose. Inflorescences of axillary or occasionally terminal spikes up to 15 cm. long; peduncle up to 6 cm. long, tomentellous. Flowers white, not sweetly scented; lower receptacle tomentose; upper receptacle pilose at the base, almost glabrous towards the apex. Fruit reddish brown, elliptic, up to 7(–9) cm. long and 3(–4·5) cm. wide, apex subtruncate and sometimes emarginate, base narrowed into the stipe, pubescent; stipe up to 15 mm. long. Cotyledons 2, borne above soil-level, with petioles 3–5 mm. long.

KENYA. Kwale District: Mrima Hill, 15 Jan. 1964, *Verdcourt* 3934!; Kilifi District: Marafa, Jan. 1937, *Dale* in *F.D.* 3649!; Lamu District: Utwani Ndogo Forest, Dec. 1956, *Rawlins* 247!

TANGANYIKA. Morogoro District: Turiani, Nov. 1954, *Semsei* 1907!; Mbeya District: Songwe, Jan. 1963, *Procter* 2360!; Lindi District: R. Mbemkuru, 6 Dec. 1955, *Milne-Redhead & Taylor* 7569!

DISTR. **K**7; **T**?1, 2, 3, 6–8; Zambia, Rhodesia, Malawi and Mozambique

HAB. Rain-forest, dry evergreen and riverine forest, also derived woodland; 70–830 m.

SYN. [*T. brownii* sensu Laws. in F.T.A. 2: 415 (1871), pro parte quoad specim. *Kirk*, non Fresen.]
　　　T. thomasii Engl. & Diels in E.M. 4: 18, t. 8/B (1900). Type: Kenya, Tana River District, Fullekullesat, *F. Thomas* 71 (B, holo. †)
　　　T. riparia Engl. & Diels in E.M. 4: 35 (1900); T.T.C.L.: 144 (1949); Griffiths in J.L.S. 55: 906 (1959). Type: Tanganyika, Iringa/Mbeya Districts, Ruaha R., *Goetze* 1457 (B, holo. †)
　　　T. aemula Diels in E.J. 39: 511 (1907); T.T.C.L.: 143 (1949). Type: Tanganyika, Lushoto District, near Amani, Sigi valley, *Engler* 3447 (B, holo. †)
　　? *T. foetens* Engl., V.E. 3(2): 724 (1921); T.T.C.L.: 145 (1949); Griffiths in J.L.S. 55: 905 (1959). Type: Tanganyika, Morogoro District, foothills of Uluguru Mts., no collector cited
　　　[*T. kilimandscharica* sensu Griffiths in J.L.S. 55: 897, fig. 27 (1959), pro parte; K.T.S.: 152 (1961), pro parte quoad specim. *Elliot* 1496, non Engl.]

NOTE. The typical form of this species, a tall forest tree in the eastern part of the Flora area, can be readily distinguished from *T. kilimandscharica*, with which it has been confused, by the additional features of the distinctly acuminate generally rather sparsely hairy leaves, but it extends up river valleys into the range of *T. kilimandscharica* and in some places it becomes difficult to differentiate without habit notes. The types of *T. hildebrandtii* Engl., P.O.A. C: 294 (1895), may well include material of both species. *Hildebrandt* 2832 (B, syn. †) from Ukamba in **K**4 was probably *T. kilimandscharica*, but *Hildebrandt* 2367 (B, syn. †) from between Duruma and Teita in **K**7 may well have been *T. sambesiaca* as the species is described as a tall tree. As both syntypes are destroyed it seems preferable to regard the name as of uncertain application and retain the later epithet *T. sambesiaca.*
　　　T. foetens and *T. thomasii* from their descriptions are also believed to belong here, but in the absence of type or authenticated specimens, it is impossible to be certain.
　　　Watkins 10, a poor specimen from **T**1, Mwanza District, Kissessa, appears to belong here, and if so forms a considerable extension of the range westwards in East Africa.

Imperfectly known species

T. bagamoyoana *Engl.,* V.E. 3(2): 724 (1921); T.T.C.L.: 144 (1949); Griffiths in J.L.S. 55: 904 (1959). Type: Tanganyika, Bagamoyo District, locality and collector not stated (B, holo. †)

Tree. Leaves not in fascicles, 10–15 cm. long, 5–6 cm. wide; venation prominent; petiole short. Receptacle ± hairy. Fruit oval, up to 3 cm. long, 2·5 cm. wide.

T. dolichocarpa *Engl. & Diels* in E.M. 4: 35 (1900); T.T.C.L.: 143 (1949); Griffiths in J.L.S. 55: 905 (1959). Type: Tanganyika, Iringa District, Ruaha R., *Goetze* 458 (B, holo. †)

Tree 10 m. high, with pyramidal lightly branched crown. Young leaves obovate-oblong, 6–10 cm. long, 2·5–3·5 cm. wide, apex somewhat obtusely acuminate, base narrowly cuneate, sericeous; lateral nerves 7–8 pairs. Fruit oblong, narrowed at each end, 9–10 cm. long, 3–3·5 cm. wide, somewhat hairy.

NOTE. Engler & Diels comment that it superficially resembles *T. sericea* DC. from which it is readily distinguished by the dimensions of the leaves and fruit.

T. morogorensis *Engl.*, V.E. 3 (2): 721 (1921); T.T.C.L.: 144 (1949); Griffiths in J.L.S. 55: 905 (1959). Type: Tanganyika, Morogoro District, precise locality and collector not cited

Tree. Leaves obovate-lanceolate; 8–10 cm. long, 4–5 cm. wide, venation slightly prominent beneath. Fruit oblong-oval to oblong, narrowed at each end, 6–7·5 cm. long, glabrous.

T. mpapwensis *Engl.*, V.E. 3 (2): 722 (1921); T.T.C.L.: 144 (1949); Griffiths in J.L.S. 55: 905 (1959). Type: Tanganyika, Mpwapwa District, Ugogo area, collector not cited.

Tree. Leaves strongly narrowed towards apex, ± hairy beneath even when mature. Fruit oblong-oval to oblong, much narrowed at each end, hairy.

T. emarginata *Engl.*, V.E. 3 (2): 723 (1921); T.T.C.L.: 145 (1949); Griffiths in J.L.S. 55: 905 (1959). Types: Tanganyika, Tabora District, Mzigwa, collector not cited & Dodoma District, Inunga, collector not cited

Leaves not fascicled, 10–15 cm. long, 5–6 cm. wide; venation prominent; petiole short. Fruit broadly oval in outline, up to 4·5 cm. long, 3 cm. wide, apex emarginate.

5. **LUMNITZERA**

Willd. in Neue Schrift. Ges. Naturf. Fr. Berlin 4: 186 (1803); Exell in Fl. Males., ser. 1, 4: 585 (1954)

Small evergreen trees or shrubs. Leaves spirally arranged, sessile or subsessile, fleshy-coriaceous. Flowers ⚥, 5-merous, regular, white, cream, yellow, pink or red, in short terminal spikes or racemes. Receptacle not externally differentiated into an upper and lower part but produced beyond the inferior ovary to form a tube bearing 2 adnate persistent bracteoles and terminating in a 5-lobed persistent calyx (or 5 sepals). Petals 5, caducous. Stamens (5)10, biseriate. Disk inconspicuous. Style filiform, persistent, not adnate to the wall of the receptacle, not expanded at the apex; ovules 2–5. Fruit indehiscent, compressed-ellipsoid and obtusely angled, ± woody, crowned by the persistent calyx. Cotyledons unknown.

Two mangrove species, one on the coast of East Africa from Kenya to Natal, Madagascar, tropical Asia, N. Australia and Polynesia, the other in tropical Asia, N. Australia and Polynesia.

L. racemosa *Willd.* in Neue Schrift. Ges. Naturf. Fr. Berlin 4: 187 (1803); Laws. in F.T.A. 2: 418 (1871); P.O.A. C: 288 (1895); Engl. & Diels in E.M. 4: 34 (1900); T.S.K., ed. 2: 36 (1936); T.T.C.L.: 142 (1949); K.T.S.: 147, fig. 29 (1961); Exell in Kirkia 7: 245 (1970). Type: India, Coromandel coast, *Klein* (B, holo., IDC microfiche *Willdenow Herb.* No. 8141 !)

Small tree or shrub up to 9 m. high; bark rough, reddish brown; young branchlets reddish or grey, sometimes appressed pubescent at first, soon glabrescent. Leaves spirally arranged; lamina narrowly obovate or narrowly obovate-elliptic to elliptic, 2–8 cm. long, 1–3 cm. wide, apex rounded, cuneate to the subsessile base or sometimes narrowed to appear subpetiolate. Inflorescences usually single, sometimes branched, axillary spikes 2–7 cm. long; rhachis glabrous. Flowers sessile or nearly so. Receptacle tubular or narrowly urceolate, laterally compressed, 6–8 mm. long, glabrous

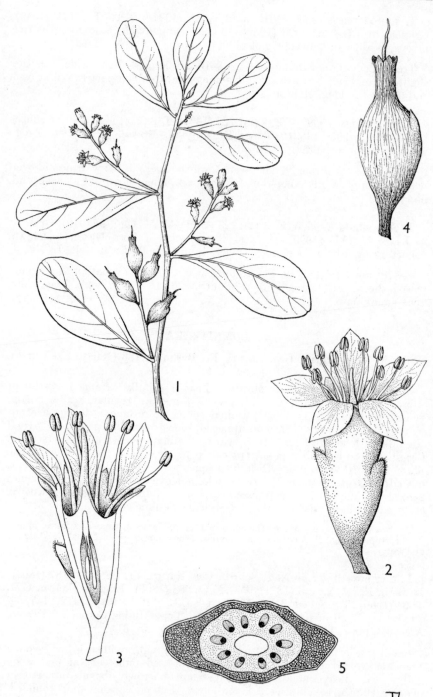

FIG. 13. *LUMNITZERA RACEMOSA*—**1**, fertile branchlet, × ⅔; **2**, flower, × 6; **3**, longitudinal section of flower, × 6; **4**, young fruit, × 4; **5**, transverse section of fruit, × 6. 1, 4, 5, from *Greenway* 4957; 2, 3, from *Vaughan* 461. Drawn by Mrs. Julia Loken. Reproduced with permission of the Editors of Flora Zambesiaca.

or pubescent, usually contracted just above the middle at the insertion of the 2 opposite or subopposite adnate bracteoles; bracteoles broadly ovate, 1·5 mm. long, sometimes ciliolate. Sepals broadly ovate-acuminate, 0·8–1 mm. long, sometimes gland-tipped. Petals white or cream (? sometimes pink) or yellow (var. *lutea* (Gaud.) Exell, confined to Timor), narrowly elliptic or narrowly obovate-elliptic, 4 mm. long, 1 mm. wide, glabrous. Stamens 10, equalling or slightly exceeding the petals. Style 6–7 mm. long, glabrous. Fruit 10–12 mm. long, 3–5 mm. wide, appressed pubescent or glabrous; pericarp with a well-developed inner layer of sclerenchyma. Fig. 13.

var. **racemosa**; Exell in Fl. Males., ser. 1, 4: 589 (1954)

Petals white or cream.

KENYA. Kilifi District: Mida, *R. M. Graham* in *F.D.* 1988! & 3 Dec. 1961, *Polhill & Paulo* 899!
TANGANYIKA. Tanga District: Machui, 2 Feb. 1967, *Perdue & Kibuwa* 8479!; Pangani District: Kumbamtoni, 25 Oct. 1955, *Tanner* 2444!; Rufiji District: Mafia I., 27 July 1932, *Schlieben* 2593!
ZANZIBAR. Zanzibar I., Mbweni, 4 Feb. 1929, *Greenway* 1323! & 7 Jan. 1930, *Vaughan* 1103! & Marahubi, 17 Apr. 1962, *Faulkner* 3033!; Pemba I., Mkoani, 8 Aug. 1929, *Vaughan* 461!
DISTR. **K**7; **T**3, 6; **Z**; **P**; Mozambique, South Africa (Natal), Madagascar, tropical Asia, N. Australia and Polynesia
HAB. Mangrove swamps, usually on the landward side at about the high water mark; sea-level to 30 m.

INDEX TO COMBRETACEAE

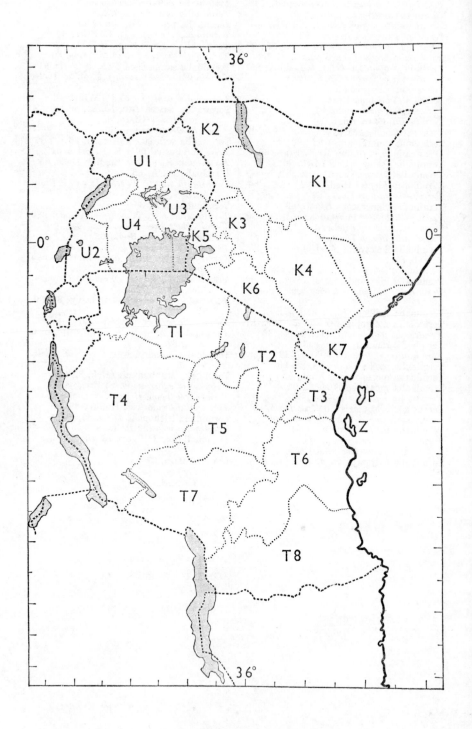